Shortcut Geometry

Colleen Schultz
and
Catherine Jeremko

New York

This publication is designed to provide accurate and authoritative information in regard to the subject matter covered. It is sold with the understanding that the publisher is not engaged in rendering legal, accounting, or other profes¬sional service. If legal advice or other expert assistance is required, the services of a competent professional should be sought.

Editorial Director: Jennifer Farthing
Editor: Cynthia Ierardo
Production Editor: Dominique Polfliet
Production Artist: Joseph Budenholzer
Cover Designer: Carly Schnur

Published by Kaplan Publishing, a division of Kaplan, Inc.
888 Seventh Ave.
New York, NY 10106

Printed in the United States of America

June 2007

10 10 9 8 7 6 5

ISBN 13: 978-1-4195-5127-7
ISBN 10: 1-4195-5127-2

TABLE OF CONTENTS

Introduction vii

Diagnostic Quiz 1

Chapter 1: Angles and Special Angle Pairs

What Are Angles and Special Angle Pairs? 27

Concepts to Help You 27

Steps You Need to Remember 33

Step-by-Step Illustration of the 5 Most Common Question Types 36

Chapter Quiz 43

Answer Explanations 45

Chapter 2: Triangles

What Are Triangles? 51

Concepts to Help You 51

Steps You Need to Remember 58

Step-by-Step Illustration of the 5 Most Common Question Types 63

Chapter Quiz 68

Answer Explanations 70

Chapter 3: Quadrilaterals and Other Polygons

What Is a Polygon? 77

Concepts to Help You 77

Steps You Need to Remember 81

Step-by-Step Illustration of the 5 Most Common Question Types 84

Chapter Quiz 88

Answer Explanations 91

Chapter 4: Circles

What Are Circles? 95

Concepts to Help You 95

Steps You Need to Remember 99
Step-by-Step Illustration of the Five Most Common Question Types . . . 103
Chapter Quiz 108
Answer Explanations 111

Chapter 5: Perimeter and Area of Polygons
What Are the Perimeter and Area of a Polygon? 117
Concepts to Help You 117
Steps You Need to Remember 120
Step-by-Step Illustration of the Five Most Common Question Types . . . 126
Chapter Quiz 131
Answer Explanations 133

Chapter 6: Surface Area
What Is Surface Area? 139
Concepts to Help You 139
Steps You Need to Remember 145
Step-by-Step Illustration of the Five Most Common Question Types . . . 147
Chapter Quiz 153
Answer Explanations 154

Chapter 7: Volume
What Is Volume? 161
Concepts to Help You 161
Steps You Need to Remember 165
Step-by-Step Illustration of the Five Most Common Question Types . . . 169
Chapter Quiz 173
Answer Explanations 175

Chapter 8: Coordinate Geometry
What Is Coordinate Geometry? 181

Concepts to Help You 181
Steps You Need to Remember 183
Step-by-Step Illustration of the Five Most Common Question Types . . . 191
Chapter Quiz 196
Answer Explanations 199

Chapter 9: Graphing Equations on the Coordinate Plane

What Are Graphed Equations? 205
Concepts to Help You 205
Steps You Need to Remember 211
Step-by-Step Illustration of the Five Most Common Question Types . . . 220
Chapter Quiz 224
Answer Explanations 227

Chapter 10: Transformational Geometry

What Is Transformational Geometry? 231
Concepts to Help You 231
Steps You Need to Remember 235
Step-by-Step Illustration of the Five Most Common Question Types . . . 242
Chapter Quiz 246
Answer Explanations 249

Chapter 11: Locus of Points

What Is a Locus of Points? 253
Concepts to Help You 253
Steps You Need to Remember 256
Step-by-Step Illustration of the Five Most Common Question Types . . . 260
Chapter Quiz 265
Answer Explanations 267

Chapter 12: Right Triangle Geometry

Why Is the Right Triangle So Important?. 273
Concepts to Help You 273
Steps You Need to Remember 279
Step-by-Step Illustration of the Five Most Common Question Types . . . 285
Chapter Quiz 290
Answer Explanations 292

Introduction

Geometry is an important part of mathematics. If you have picked up this book, maybe this was a part of your math instruction that has always been a challenge for you. You may be a high school or college student learning this material for the first time. You may be using this book to refresh and relearn important topics in preparation for a standardized test. You may be a professional and have a need to apply geometric principles to some aspect of your work.

Whatever your need, this book is designed for you! *Shortcut Geometry* is a concise, easy-to-understand method for learning or reviewing the world of two- and three-dimensional figures. This book will provide you with over 200 sample problems that you would commonly encounter with geometry. Once you have worked through the numerous examples and problems provided with answer explanations, you will feel more confident in your ability to tackle challenges.

In addition to providing a refresher on the many topics that encompass this branch of mathematics, this book will guide you through the various forms that a question can take. For example, consider these:

- What is the volume of a cylinder with a radius of 4 cm and a height of 10 cm?
- What is the volume of a cylinder with a diameter of 8 cm and a height of 10 cm?
- If the volume of a cylinder is 160π cm^3, what is the radius?
- If the volume of a cylinder is 160π cm^3, what is the diameter?

All of the above are just different instances of the same problem. In order to solve each of these, you will use a formula; in each of the types, you use the formula to solve for different quantities.

Shortcut Geometry begins with the basic building blocks of points, lines, rays, and angles. From there, the book will cover the basic polygons, the measurements of perimeter, area, surface area, and volume, and then move on to the other concepts that encompass this branch of math.

To get the most out of the content of this book, start your studies off with the Diagnostic Test. Following the test are the answers, and then the detailed answer explanations. After the explanations, there is a correlation chart that will guide you to the chapter in the book that details the concepts covered by any particular question on the test.

If time permits, we recommend that you work through the entire book. Even if you answered the diagnostic question correctly, you will benefit from studying the chapter that covers that particular topic. Not every concept, skill, and technique can be represented in the pretest. The diagnostic is a way to measure your prior knowledge, to celebrate your strengths, and to steer you towards any areas of weakness. It is not meant to represent the entire content of each chapter.

Each of the 12 chapters is structured to identify key concepts and to outline steps that will help you solve the most common types of problems. The text is sprinkled throughout with sample questions and detailed explanations that give you effective techniques and strategies. Each chapter concludes with a brief quiz to apply your knowledge and to evaluate your growth.

After reading and working through *Shortcut Geometry*, we're confident that you will have a better understanding of the essential concepts. You'll be amazed at what you can accomplish in so little time! So let's get started on mastering geometry!

Good luck—and enjoy the shortcut!

Diagnostic Test

1. Two complementary angles are in the ratio 1:5. What is the measure in degrees of the larger angle?

   (A) 15°

   (B) 18°

   (C) 60°

   (D) 75°

   (E) 90°

2. What is the measure in degrees of angle ∠ABC in the diagram below?

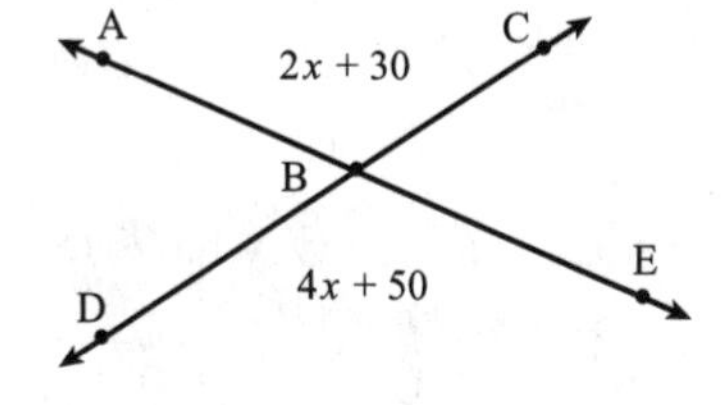

   (A) 33.3°

   (B) 40°

   (C) 80°

   (D) 96°

   (E) 110°

3. In the figure below, two parallel lines l and m are cut by transversal n. What is the value of x?

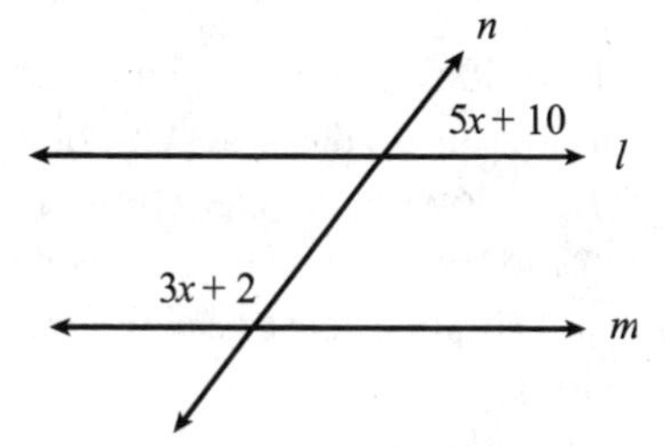

   (A) 21

   (B) 24

   (C) 65

   (D) 84

   (E) 115

4. Which of the following choices best describe the triangle in the figure below?

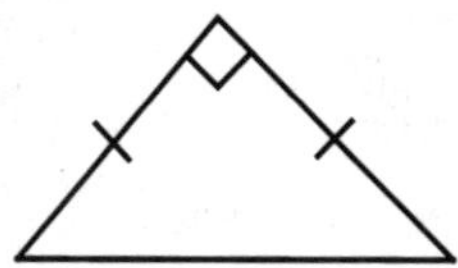

   (A) Acute scalene

   (B) Right scalene

   (C) Right isosceles

   (D) Acute isosceles

   (E) Obtuse scalene

5. Find the value of x in the figure below.

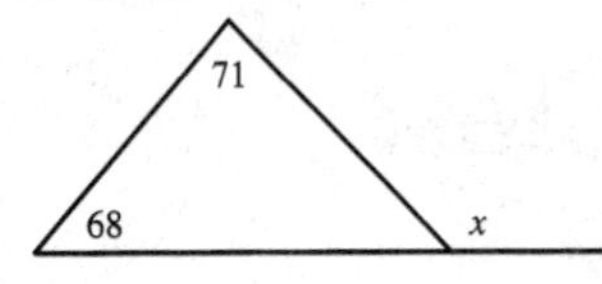

(A) 41
(B) 139
(C) 141
(D) 149
(E) 180

6. Two sides of a triangle measure 4 units and 10 units. Which of the following answer choices could represent the number of units in the measure of the third side?

(A) 4
(B) 5
(C) 6
(D) 10
(E) 15

7. What is the name of a six-sided polygon?

(A) Nonagon
(B) Octagon
(C) Hexagon
(D) Quadrilateral
(E) Pentagon

8. What is the degree measure of an interior angle in a regular pentagon?

(A) 50°
(B) 90°
(C) 108°
(D) 180°
(E) 540°

9. In the rectangle ABCD shown below, what is the length of side BC?

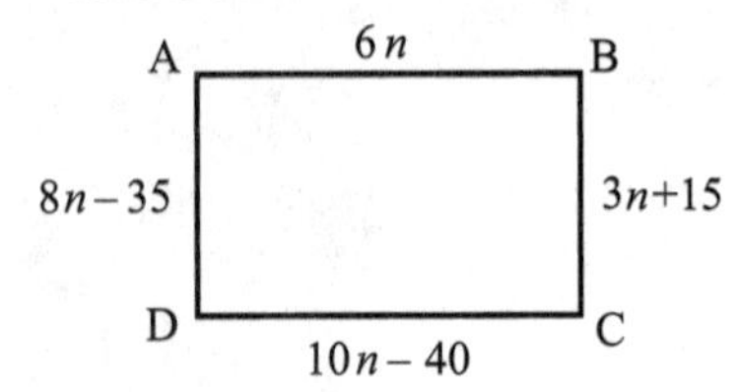

(A) 5 units
(B) 10 units
(C) 30 units
(D) 45 units
(E) 60 units

10. In the circle below, what is the measure of chord BD?

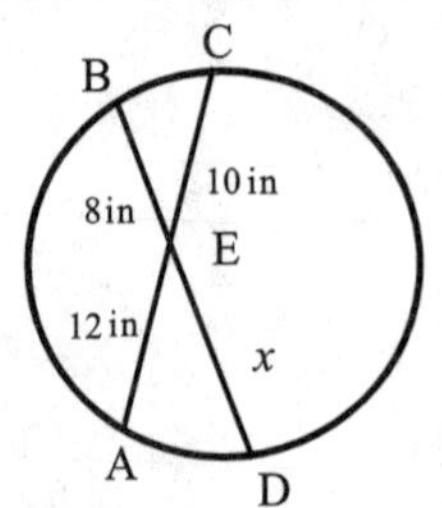

(A) 9.6 in
(B) 10 in
(C) 15 in
(D) 18 in
(E) 23 in

11. Given the following circle with tangent and secant as shown, what is the measure of ∠ABC?

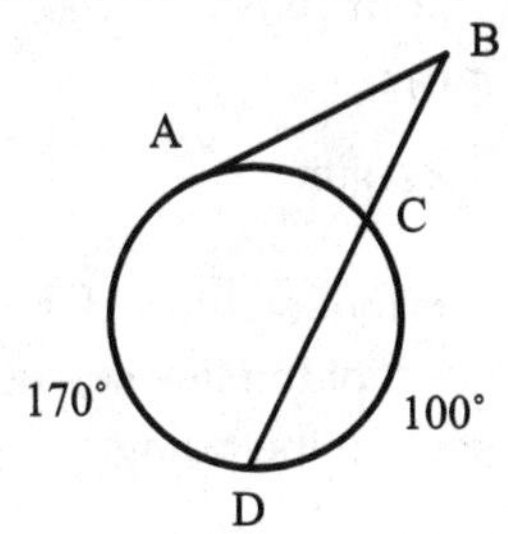

(A) 40°
(B) 80°
(C) 90°
(D) 130°
(E) 135°

12. What is the perimeter of the following figure?

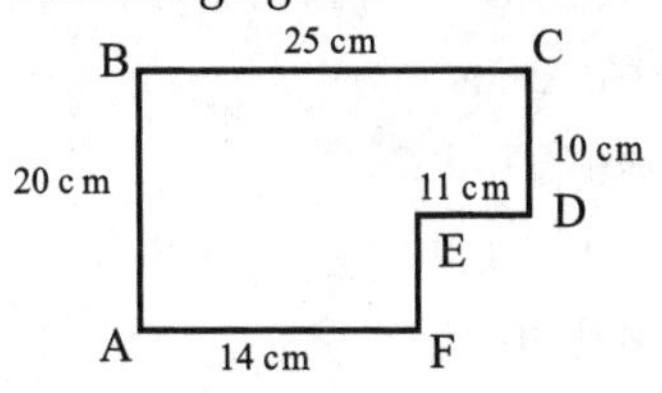

(A) 59 cm
(B) 80 cm
(C) 90 cm
(D) 170 cm
(E) 280 cm

13. If the area of the trapezoid shown below is 100 mm^2, what is the length of the base CD?

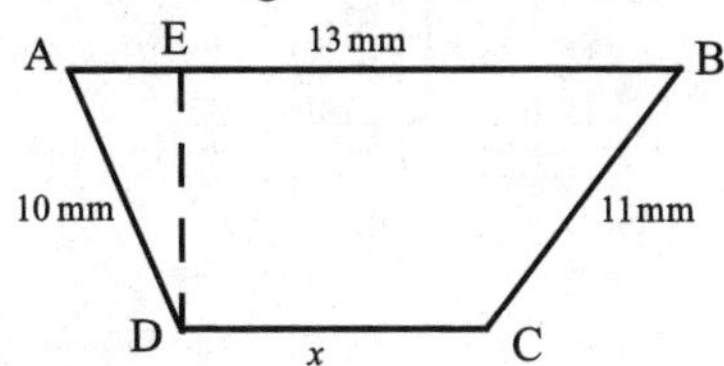

(A) 7 mm
(B) 8.7 mm
(C) 23 mm
(D) 43 mm
(E) 130 mm

14. What is the area of the shaded region in the figure below? Leave your answer in terms of π.

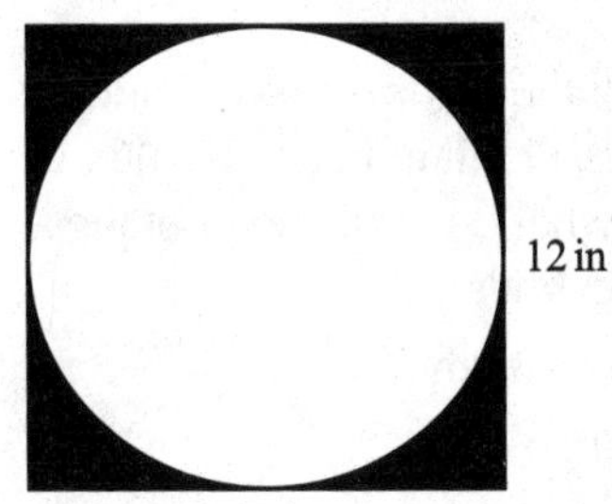

(A) $144 - 12\pi$ in^2
(B) $144 - 36\pi$ in^2
(C) 36π in^2
(D) 144 in^2
(E) $144\pi - 144$ in^2

15. How many square units are there in the surface area of a cube whose edge measures 4 cm?

(A) 12 cm^2
(B) 16 cm^2
(C) 48 cm^2
(D) 64 cm^2
(E) 96 cm^2

16. What is the surface area, in terms of π, of a cylinder with a base diameter of 10 m and a height of 6 m?

(A) 50π m

(B) 60π m

(C) 110π m

(D) 150π m

(E) 260π m

17. If the volume of a cylinder that is 12 cm in height is 108π cm^3, what is the diameter of the cylinder?

(A) 3 cm

(B) 6 cm

(C) 9 cm

(D) 18 cm

(E) 28.26 cm

18. In the rectangular pyramid below, the height (the dotted-line segment) is 10 inches. What is the volume?

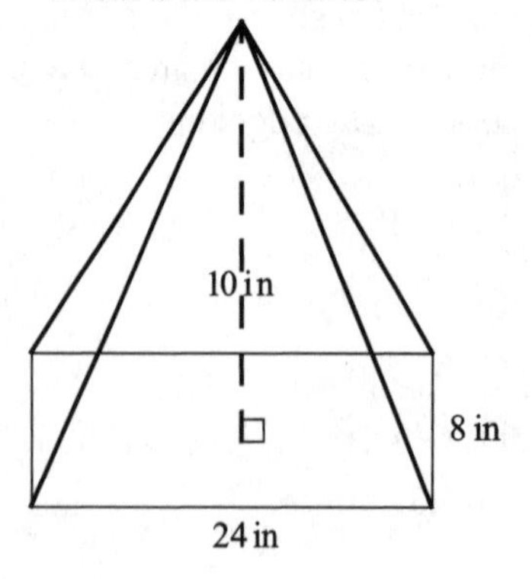

(A) 42 in^3

(B) 80 in^3

(C) 192 in^3

(D) 640 in^3

(E) 1,920 in^3

19. What is the location of the midpoint of the line segment connecting the points (–2, 3) and (4, –9)?

(A) (1, –3)

(B) (1, –6)

(C) (2, –6)

(D) (–1, 3)

(E) (–1, –3)

20. What is the distance between the points (1, 2) and (4, 6)?

(A) units

(B) 5 units

(C) 7 units

(D) units

(E) None of these

21. What is the equation of the line in the figure below?

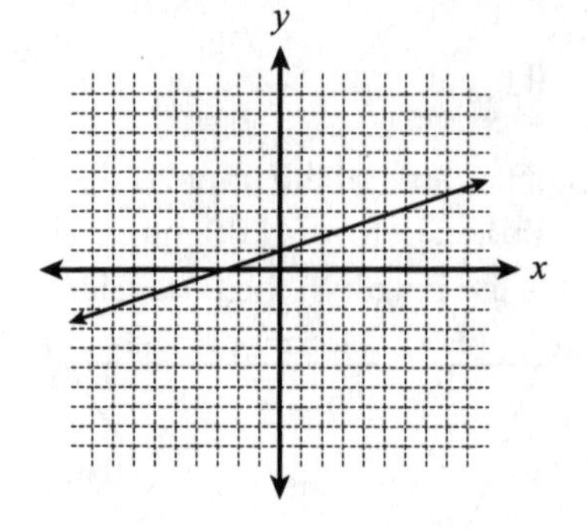

(A) $y=\frac{1}{3}x-1$

(B) $y=-\frac{1}{3}x-1$

(C) $y=3x-1$

(D) $y=-3x+1$

(E) $y=\frac{1}{3}x+1$

22. What is the equation of the circle shown below?

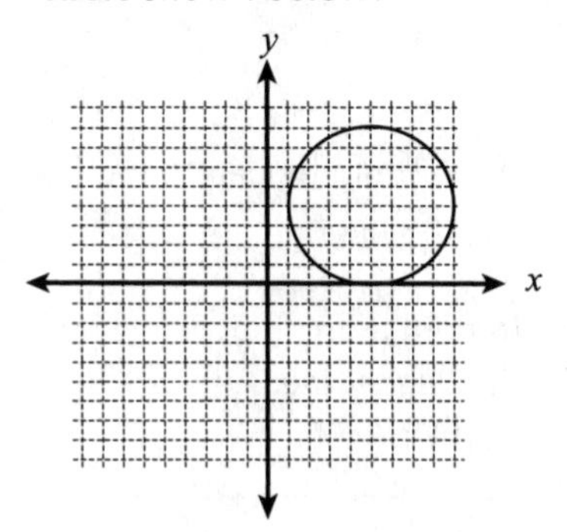

(A) $(x-5)^2+(y-4)^2=4$

(B) $(x-5)^2+(y-4)^2=16$

(C) $(x^2+5)+(y^2+4)=4$

(D) $(x^2-5)+(y^2-4)=16$

(E) $(x+5)^2+(y+4)^2=16$

23. What are the coordinates of the vertex of the equation $y = 3x^2 - 12x + 6$?

(A) (–2, –6)

(B) (0, 6)

(C) (2, –6)

(D) (–4, 6)

(E) (4, –12)

24. Which of the following is the number of lines of symmetry in a regular pentagon?

(A) 1

(B) 2

(C) 5

(D) 10

(E) None

25. Which choice best describes the transformation in the figure below?

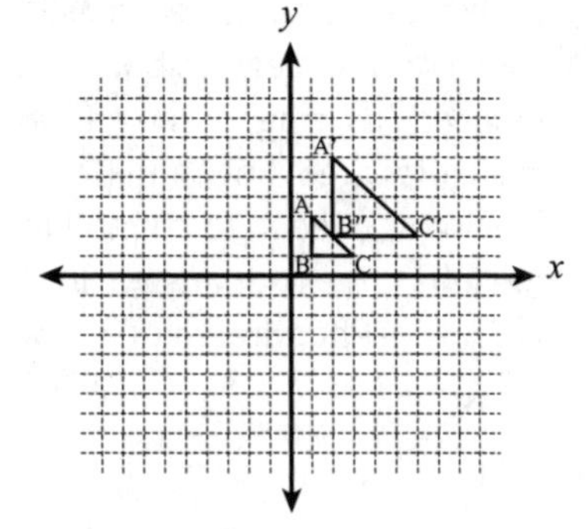

(A) Translation

(B) Reflection

(C) Dilation

(D) Rotation

(E) Isometry

26. How many points are contained in the intersection of the locus of points 3 units from a given line l and 5 units from a given point P located on line l?

(A) 0

(B) 1

(C) 2

(D) 4

(E) 6

27. Which of the following equations best describes the locus of points 6 units from the origin?
    (A) $x^2 + y^2 = 6$
    (B) $x^2 + y^2 = 36$
    (C) $y = x^2 + 36$
    (D) $x = y^2 + 36$
    (E) $y = x^2 + 6$

28. Miguel wants to enclose a triangular area in his backyard for his dog to roam freely, as shown in the diagram below. If he needs fencing for the entire triangular area, how much fencing is needed to co mplete the job? Give your answer to the nearest hundredth of a foot.

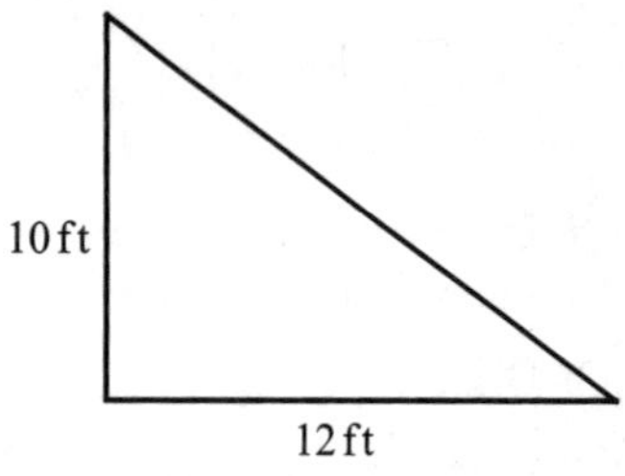

    (A) 6.63 ft
    (B) 15.62 ft
    (C) 22 ft
    (D) 28.63 ft
    (E) 37.62 ft

29. Given right triangle ΔLMN, and length NL is 24 m, what is the length of NM, to the nearest hundredth of a meter?

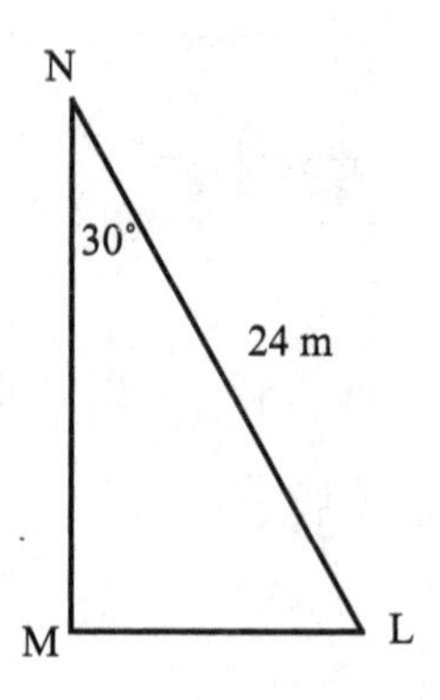

    (A) 12 m
    (B) 20.78 m
    (C) 24 m
    (D) 36 m
    (E) Cannot be determined from the figure

30. In right triangle ΔCDE, what is the degree measure of angle ∠DEC?

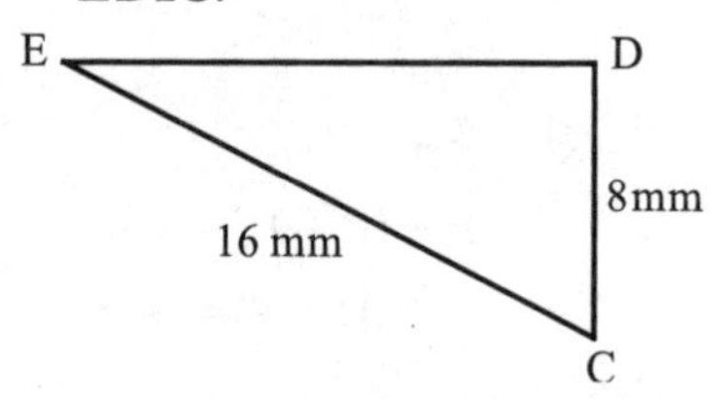

    (A) 15°
    (B) 30°
    (C) 45°
    (D) 90°
    (E) Cannot be determined from the figure

DIAGNOSTIC TEST ANSWERS

1. D	16. C
2. E	17. B
3. A	18. D
4. C	19. A
5. B	20. B
6. D	21. E
7. C	22. B
8. C	23. C
9. D	24. C
10. E	25. C
11. A	26. D
12. C	27. B
13. A	28. E
14. B	29. B
15. E	30. B

EXPLANATIONS

1. D

Properties of special angle pairs are discussed in Chapter 1. In this case, the special angle relationship is a pair of complementary angles. The sum of any two complementary *angles* is always 90 degrees.

In this question you are also given the ratio between the two angles. Use these ratios to help you write an equation to solve. Since the ratio of the two angles is 1:5, represent the smaller angle as $1x$ and the larger angle as $5x$. Next, set up the equation $1x + 5x = 90$, since the sum of the two angles is 90 degrees.

Start with the equation and look for like terms: $1x + 5x = 90$

Since $1x + 5x = 6x$, the equation becomes: $6x = 90$

Divide each side of the equation by 6 to get the x alone: $\frac{6x}{6} = \frac{90}{6}$

$x = 15$

The value of $x = 15$. Since the larger angle is represented by $5x$, multiply $5(15) = 75$ to find the measure of this angle.

Choice (A) is the value of x in the question and would represent the smallest angle, but is not the measure of the largest angle. Choice (B) would be the result of dividing 90 by 5, instead of 6, to find x. Choice (C) is the result of multiplying the value of x by 4, in the case that the two numbers in the ratio were subtracted instead of added. Choice (E) is incorrect because it represents the sum of the two angles, not just the measure of the larger angle.

2. E

Vertical angles are a special type of angle pair. Vertical angles are the nonadjacent angles formed by two intersecting lines. This concept is explained in Chapter 1. The two angles labeled in the diagram are vertical angles and will have the same measure. Set the two expressions equal to each other to first find x and then find the measure of angle $\angle ABC$.

Start by setting the two expressions equal to each other:	$2x + 30 = 4x - 50$
Subtract $2x$ from each side of the equation:	$2x - 2x + 30 = 4x - 2x - 50$
This simplifies to the equation:	$30 = 2x - 50$
Add 50 to each side of the equation:	$30 + 50 = 2x - 50 + 50$
The equation becomes:	$80 = 2x$
Divide each side of the equation by 2 to find x.	$\frac{2x}{2} = \frac{80}{2}$
	$x = 40$

Substitute the value of x into the expression $2x + 30$ to find the measure of angle$\angle ABC$.

$2(40) + 30 = 80 + 30 = 110$ degrees.

Choice (A) would be the value of x if the sum of the expressions for two angles were incorrectly set equal to 180, and choice (D) would be the measure of the angle in this instance. Choice (B) is the value of x in this question, but is not the measure of angle $\angle ABC$. Choice (C) is the measure of the other two vertical angles in the figure, angles $\angle ABD$ and $\angle CBE$.

3. A

There are a variety of situations that emerge when two parallel lines are cut by a transversal, and these situations are explained in Chapter 1. The first step to solve a question with the angles formed by parallel lines and a transversal is to decide whether the two angles are congruent or supplementary. In this case, the angle labeled $5x + 10$ is on the exterior of the parallel lines to the right of the transversal and would be congruent to the angle labeled 1 in the figure below, because they are corresponding angles.

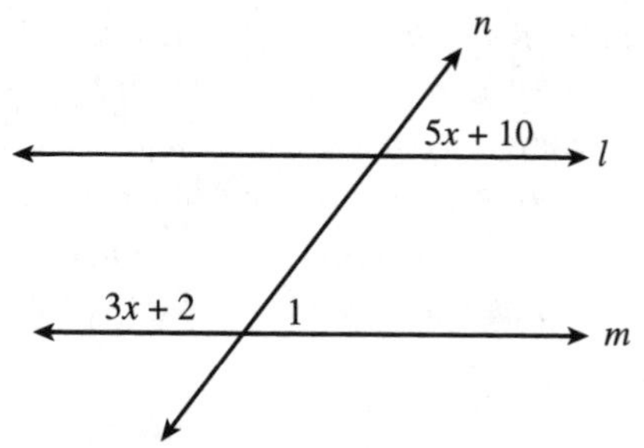

That angle is adjacent to the angle labeled $3x + 2$, and the two angles form a straight angle. Therefore, that angle would be supplementary to the angle labeled $3x + 2$.

To solve for x, add the two expressions and set them equal to 180 degrees.

The equation would be: $5x + 10 + 3x + 2 = 180$
Combine like terms to get: $8x + 12 = 180$
Subtract 12 from both sides of the equation: $8x + 12 - 12 = 180 - 12.$
$8x = 168$

Divide each side of the equation by 8 to get x alone. $\frac{8x}{8} = \frac{168}{8}$

Since 168 divided by 8 is 21, $x = 21$.

Choice (C) is the measure of the angle labeled $3x + 2$, and choice (E) is the measure of the angle labeled $5x + 10$. Even though these would be the correct measures, recall that the question only asked for the value of x, not for the measure of either angle.

4. C

The figure is an isosceles right triangle. Triangles are classified, or named, based on their sides and their angles. These different types of triangles are explained in Chapter 2. If a triangle has three sides that all have the same measure, it is called an *equilateral triangle.* If only two sides have the same measure, the triangle is called an *isosceles triangle.* If all of the sides have different measures, the triangle is *scalene.* In this triangle, two sides are congruent; therefore, the triangle is isosceles.

The type of angles it contains also classifies a triangle. If it has all acute angles, or all angles have less than 90 degrees, the triangle is an *acute triangle.* If it contains one angle greater than 90 degrees, or one obtuse angle, the triangle is an *obtuse triangle.* If the triangle contains a right angle (an angle that measures 90 degrees), the triangle is known as a *right triangle.* This triangle contains a right angle, as well as two congruent sides, making it an isosceles right triangle.

5. B

The missing exterior angle is 139 degrees. The procedure to find a missing exterior angle of a triangle is explained in Chapter 2. Extending one side of a triangle creates an exterior angle. The angle is located between this extended side and an adjacent side of the triangle. One way to calculate this measure is to find the sum of the two interior angles that are not adjacent to the angle labeled x. These are called the two remote interior angles. This sum is equal to the measure of the exterior angle x.

To do this, first locate the angle labeled x. Then, look for the two angles in the interior of the triangle that are the farthest away from this angle. These two angles are labeled 71 and 68 degrees in the figure. Since $71 + 68 = 139$, then the measure of the exterior angle is also 139. Thus, $x = 139$.

Another way to find this value is to first determine the measure of the interior angle that is adjacent to the exterior angle x. Then subtract this value from 180, since these two angles form a straight angle. The sum of the degrees in the interior of a triangle is 180 degrees. The interior angle is equal to 41, since $41 + 71 + 68 = 180$. Next, subtract this angle measure from 180 degrees since this angle and angle x form a straight angle. The measure of x is $180 - 41 = 139$.

Choice (A) is incorrect and would be the number of degrees in the interior angle adjacent to the angle labeled x. Choice (C) could be the result of not subtracting correctly. Choice (E) is the sum of the three interior angles, or the sum of the exterior angle and the adjacent interior angle.

6. D

Various properties of triangles are covered in Chapter 2. One of these properties is the *Triangle Inequality Property.* This property states that the sum of the two smaller sides of a triangle must be greater than the third side.

To find possible measures for the third side, this measure cannot be greater than or equal to the sum of the two given sides, and it cannot be less than or equal to the difference between the two given sides.

The two given sides are 4 and 10. Find the sum and the difference between these two numbers in order to find the two values in which the third side is between. $10 - 4 = 6$ and $10 + 4 = 14$. Therefore, the third side must be between, but not include, 6 and 14. The only value that fits this criterion is choice (D) 10.

Choices (A), (B), and (C) are values that are too small for the measure of the third side and choice (D) would be too large.

7. C

This question tests your knowledge of classifying polygons. This will be covered in Chapter 3. A polygon is a closed figure made up of line segments, called *sides.* Polygons are named according to the number of sides they contain. A six-sided polygon is called a *hexagon.* Choice (A), a *nonagon,* is a nine-sided figure. An *octagon,* choice (B) has eight sides. A *quadrilateral,* choice (D), has four sides and is one of the most common polygons used in geometry. Choice (E), a *pentagon,* has five sides.

8. C

You need to be familiar with several geometric definitions and facts about polygons to answer this question, which is covered in Chapter 3. An *interior angle* is one of the angles enclosed within a polygon. If you did not remember what a regular polygon was, you might wonder which angle the question refers to. A regular polygon is one in which all sides and all angles are congruent. A pentagon has five sides and five angles. The last fact that you need is that the sum of the interior angles of a regular polygon is equal to $180(n-2)$ degrees, where n is equal to the number of sides. The sum of all of the interior angles in this 5 sided polygon is $180(5-2) = 180(3) = 540°$. So one of the interior angles would be $540 \div 5 = 108°$.

If you chose choice (B), you may have been thinking of a regular quadrilateral. If your choice was (E), you gave the sum of all the angles, not the measure of one of the angles.

9. D

Using algebra to solve quadrilateral problems is described in Chapter 3. You are told the figure is a rectangle. A *rectangle* is a parallelogram, which has opposite sides that are equal in length. To find the length of side BC, set the expression represented by BC, $3n + 15$, equal to the expression denoted by side AD, $8n - 35$. Then solve the equation: $8n - 35 = 3n + 15$. This is an equation with a variable on both sides.

First, subtract $3n$ from both sides of the equation: $8n - 3n - 35 = 3n - 3n + 15$

Combine like terms to get: $5n - 35 = 15$

Add 35 to both sides of the equation: $5n - 35 + 35 = 15 + 35$

Again, combine like terms: $5n = 50$

Divide both sides by 5: $\frac{5n}{5} = \frac{50}{5}$, or $n = 10$.

Now that you know the value of n, substitute in this value of 10 into the expression for BC, that is $3n + 15$, to get $3(10) + 15 = 45$ units.

If your choice was (A) you may have thought the figure was a square, and incorrectly set side AB equal to side BC, and solved the equation $6n = 3n + 15$. If you did this, you would have arrived at $n = 5$. Choice (B) is the value of n, not the length of side BC. Choice (C) could have been arrived at by solving the incorrect equation as described in choice (A), and then substituting in the value of 5 for n in the expression $3n + 15$.

10. E

When two chords intersect in a circle, the product of the segment parts of one chord is equal to the product of the segment parts of the other chord. Many facts about circles are covered in Chapter 4. Notice in the figure that there are two segments within the circle that make up chord AC, AE and EC. Now look at chord BD. It is made up of two segments, BE and ED. The product of segments AE and EC is equal to the product of segments BE and ED. Set up this equation as (AE)(EC) = (BE)(ED).

Now, substitute in the given measures to get $(12)(10) = 8x$, or $8x = 120$.

Divide both sides of this equation by 8 to get $\frac{8x}{8} = \frac{120}{8}$, or $x = 15$.

If your answer was choice (C), you did not finish the problem. You are asked to find the length of chord BD, not just the value of x. The length of chord BD is $8 + x = 8 + 15 = 23$ inches, choice (E).

If your choice was (B) or (D), you may have made an incorrect assumption; you looked and saw that the segments of chord AC differed by 2, and thought that $x = 10$, 2 units longer than BE. This is not true of intersecting chords in a circle.

11. A

This figure shows a tangent AB and secant BD that intersect at B to form angle ∠ABC. There is a formula to calculate the measure of this angle. This, and other formulas to find angle and segment measures related to circles are described in Chapter 4. The measure of angle ∠ABC is one-half of the difference between the intercepting arcs, AD and AC. You need to determine the measures of arc AD and arc AC. The measure of angle ∠ABC is $\frac{1}{2}(AD - AC)$. The measure of arc AD is given as 170 degrees. The measure of arc AC is not given. Use the fact that a circle has a total degree measure of 360 to find the measure of arc AC: AC = 360 – (170 + 100), or AC = 360 – 270 = 90 degrees. Now use the two arc measures in the formula. $\angle ABC = \frac{1}{2}(170 - 90) = \frac{1}{2}(80) = 40°$

If your choice was (B), you most likely forgot to take one-half of the difference. Answer choice (C) is the measure of arc AC. If your answer was (D), you took one-half of the sum, instead of the difference, to find the measure of ∠ABC.

12. C

Perimeter is the distance around a figure. To find information about perimeter and area, look in Chapter 5. To determine the perimeter, add up all of the side lengths around the figure. The measure of side EF is not given. But notice that side CD + EF = AB. Use this observation to calculate that EF = AB – CD, or EF = 20 – 10 = 10 cm. Now add all of the sides to get the perimeter. AB + BC + CD + DE + EF + FA, or 20 + 25 + 10 + 11 + 10 + 14 = 90 cm.

Answer choice (A) is the sum of the three longest sides. Choice (B) is the sum of all of the sides whose measures are listed on the figure. If you chose either choice (D) or (E), you were probably thinking about the area of the figure. Choice (E) is the area of a rectangle that is 20 long and 14 wide. Choice (D) is the area of the irregular figure shown.

13. A

The area of polygons, including trapezoids, is covered in Chapter 5. Area is the amount of square units it takes to cover a polygon. The formula for the area of a trapezoid is $A = \frac{1}{2}h(b_1 + b_2)$, where h is the height and b_1 and b_2 are the parallel bases.

In this problem, you are given the area of 100 mm^2; the height is the dotted line in the figure, of length 10 mm. One of the measures of the bases, AB, is shown to be 13 mm. Let x represent the length of base CD; substitute in all known values in the formula to get $100 = \frac{1}{2} \bullet 10 \bullet (x + 13)$.

Solve this equation by first multiplying: $\frac{1}{2} \bullet 10 = 5$

The equation becomes: $100 = 5(x + 13)$

You can now divide both sides of the equation by 5: $\frac{100}{5} = \frac{5(x+13)}{5}$

The equation becomes: $20 = x + 13$

Isolate the variable x by subtracting 13 from both sides: $20 - 13 = x + 13 - 13$

The length of CD: $7 = x$

Answer choice (C) is the sum of one of the bases and the height. If your answer was choice (D), you thought of area and just multiplied base times height. In choice (E), you may have incorrectly thought this was a rectangle and multiplied 13 • 11, and subtracted 100 from this product.

14. B

Finding the area of a shaded region is a common question type that is often encountered on various types of tests. Area of shaded regions is described in Chapter 5. To find the area of the shaded region, notice that the outer figure is a square, because a circle is inscribed within it. You can think of the shaded region as actually the area of a square with the circular area removed. So the area of the shaded region is $Area_{square} - Area_{circle}$. The area of a square is $A = s^2$, where s is the length of a side of the square. The area of a circle is $A = \pi r^2$, where r is the radius of the circle. At first glance, it seems that you cannot determine the radius of the circle. But, since the circle is inscribed in the square, the diameter of the circle is the same length as the side of the square, that is 12 m. The radius, r, is one-half the diameter, so the radius is 6 m. The area of this shaded region is $s^2 - \pi r^2$, where $s = 12$ and $r = 6$. The shaded area is $12^2 - \pi 6^2$. Follow order of operations, and evaluate the exponents first to get $144 - 36\pi$.

If your answer was (A), you multiplied 6 • 2 instead of 6 • 6 when evaluating r^2. Choice (C) is the area of the circle, and choice (D) the area of the square. In choice (E), the radius was mistakenly thought to be 12 instead of one-half of the diameter, which is 6.

15. E

The surface area of a figure is the area of its faces, or what you might think is the outside covering of a three-dimensional object. Surface area, and the various formulas that are used for different objects, are covered in Chapter 6. When finding surface area you are working in two dimensions, so be sure to use square units.

This question asks for the surface area of a cube. When picturing a cube, think of dice from a board game. Each cube has 6 faces. Each face of a cube is congruent to every other face. In addition, each face is a square which has four congruent sides, which are called *edges* in three-dimensional objects. Therefore, the area of each can be found by multiplying *edge* • *edge*, or e^2.

Since there are 6 congruent faces, the formula becomes $SA = 6 \bullet e^2$ or $6e^2$.

For this question, the length of an edge of the cube is 4 cm. Substitute this value into the surface area equation. $SA = 6e^2$

After substituting, the equation becomes: $SA = 6(4)^2$

Follow the correct order of operations and apply the exponent first: $SA = 6(16)$

Then multiply: $SA = 96 \text{ cm}^2$

Choice (A) is the result of multiplying the length of the edge by 3. Choice (B) is the area of one face of the cube, but since you are looking for the total surface area, this value needs to be multiplied by 6. Choice (C) is the result of multiplying 16 by 3, instead of 6. You may have selected choice (D) if you had used the formula for the volume of the figure ($V = e^3$), instead of the surface area.

16. C

The concept of surface area, covered in Chapter 6, is the area of the outside covering of a three-dimensional object, or solid. Finding the surface area of a cylinder is a little more complicated than some other solids because it has curved surfaces. However, this question asks for the answer to be left in terms of π, so keep that symbol in your answer and do not use an approximation for this value.

The outside covering of a cylinder is made up of two circles (for the bases) and a rectangular area that wraps around the height of the cylinder. When picturing this rectangular area, think of a label on a soup can. Even though it wraps around in a circular fashion, when the label is removed and laid flat, it is in the shape of a rectangle. The formula for finding the surface area of a cylinder is $SA = 2\pi r^2 + \pi dh$, where r = radius of the base, d = diameter of the base, and h = height of the cylinder. The $2\pi r^2$ in the formula represents the area of the circular bases and the πdh represents the area of the rectangular area that wraps around the base.

To find the surface area, substitute into the formula and simplify: $SA = 2 r^2 + \pi \text{ dh}$

Substitute $d = 10$, $r = 5$ (half of the diameter), and $h = 6$ into the formula: $SA = 2 \pi (5)^2 + \pi (10)(6)$

Perform the correct order of operations by applying the exponent first:	$SA = 2\pi(25) + \pi(10)(6)$
Now, multiply within each term of the equation:	$SA = 50\pi + 60$
Combine the values to get the surface area:	$SA = 110 \text{ m}^2$

Choice (A) is the surface area of the two circular bases, but does not include the surface area of the rectangle that wraps around the height of the cylinder. Choice (B) is the result of multiplying the diameter of the figure by the height. If you mistakenly found the volume of the figure ($V = \pi r^2h$), you may have selected choice (D).

17. B

Volume is the number of cubic units it takes to fill a three-dimensional figure. You can learn more about volume in Chapter 7. Generally speaking, the volume of a prism or a cylinder is the area of the base times the height. A cylinder has two circular bases that are a certain distance apart, which is the height. The formula for the volume of a cylinder is $V = \pi r^2h$. In this problem, you are given the height and the volume of a cylinder, and you are asked to find the diameter. To answer this question, you will first have to solve for the radius of the cylinder and then double that value to find the diameter. Substitute the known values into the formula to get $108\pi = \pi r^2(12)$.

Divide both sides of this equation by 12π to get: $\frac{108\pi}{12\pi} = \frac{12\pi r^2}{12\pi}$, or $9 = r^2$.

To solve for the radius, take the square root of each side and $r = 3$.

The radius is one-half the length of the diameter, so the diameter is $3 \cdot 2 = 6$ cm.

Answer choice (A) is the radius, not the diameter. Choice (C) is the value of the radius squared, and choice (D) is two times the incorrect value 9. If you chose choice (E), you most likely just divided 108π by 12.

18. D

The volume of pyramids is covered in Chapter 7. Volume is the number of cubic units it takes to fill a three dimensional figure. A rectangular pyramid's volume is one-third the volume of a rectangular solid. The formula for a rectangular solid is $V = lwh$, where l is the length, w is the width, and h is the height of the figure. Therefore, the formula for the volume of this pyramid is $\frac{1}{3}lwh$. Substitute in the given dimensions to get $V = \frac{1}{3} \bullet 10 \bullet 8 \bullet 24 = \frac{1}{3} \bullet 1{,}920 = 640 \text{ in}^3$.

If you chose answer choice (A), you added the given measures. Choice (C) is the area of the base rectangle, not the volume of the pyramid. Choice (E) is the volume of the rectangular prism with the same dimensions. The pyramid is one-third as big.

19. A

The *midpoint* between two points in the coordinate plane is the point halfway from one point or another. In order to find each coordinate x and y, find the average of the two x values and the average of the two y values. The average for each is found by adding the two values for each and dividing this sum by 2. Therefore, the formula for midpoint is $\left(\frac{x_1 + x_2}{2}, \frac{y_1 + y_2}{2}\right)$. This concept is further explained in Chapter 8.

To find the midpoint between the two points given in the question, take the two points (–2, 3) and (4, –9) and use them as (x_1, y_1) and (x_2, y_2). Then, substitute those values into the formula $\left(\frac{x_1 + x_2}{2}, \frac{y_1 + y_2}{2}\right)$.

After substituting, the formula becomes:

$$\left(\frac{x_1 + x_2}{2}, \frac{y_1 + y_2}{2}\right) = \left(\frac{-2+4}{2}, \frac{3+-9}{2}\right)$$

Now, simplify the values in parentheses to find the coordinates.

Add the numerators of each fraction: $\left(\frac{2}{2}, \frac{-6}{2}\right)$

Finally, simplify each fraction to find the midpoint. Since $2 \div 2 = 1$ and $-6 \div 2 = -3$, the midpoint is (1, –3).

Choices (B) and (D) are each the result of computational errors with the negative numbers. Choice (D) and choice (E) are the result of confusing the signs in the problem.

20. B

The distance between two points in the coordinate plane can be found by using the distance formula. This formula is based on the Pythagorean Theorem and is equal to $d = \sqrt{(x_1 - x_2)^2 + (y_1 - y_2)^2}$. This topic is covered in Chapter 8.

To apply this formula, take the two given points in the problem (1, 2) and (4, 6) and use them as (x_1, y_1) and (x_2, y_2).

To find the distance between any two points, use the *distance formula*.

$$d = \sqrt{(x_1 - x_2)^2 + (y_1 - y_2)^2}$$

Substitute the given points into the distance formula:

$$d = \sqrt{(1-4)^2 + (2-6)^2}$$

Combine the values within the parentheses.
Square each of the numbers in parentheses.
Recall that $(-3)^2 = (-3)(-3) = 9$.

$$= \sqrt{9+16}$$

Add the values under the square root symbol:

$$= \sqrt{25}$$

$$= 5$$

The square root of 25 is equal to 5, since 5 • 5 = 25. Therefore, the distance is 5 units.

Answer choice (A) is the result of simply adding 3 + 4 = 7 under the square root sign, and not squaring each number first. Choice (C) would result after the same error was made, and the number was not kept under the square root sign. You may have selected choice (D) if you had forgotten to get rid of the square root sign after the square root of 25 was evaluated.

21. E

Coordinate geometry and writing the equations of lines is covered in Chapter 8. Since the figure contains a straight line, the equation is linear and in the form $y = mx + b$. To find the equation of the line in the figure, first find the slope and y-intercept of the line. Then, use the equation $y = mx + b$ and replace the slope for m and the y-intercept for b. Use the fact that lines that have positive slope slant up to the right and lines that have negative slope slant up to the left to help you with this type of question. The line intersects the y-axis at the point (0, 1), so the y-intercept is 1. The line goes through the points (0, 1) and (3, 2) so use the slope formula to find the slope of the line.

The *slope formula* is $m = \frac{y_1 - y_2}{x_1 - x_2}$.

Take the two points (0, 1) and (3, 2) and use them as (x_1, y_1) and (x_2, y_2).

$$m = \frac{1-2}{0-3} = \frac{-1}{-3} = \frac{1}{3}$$

In addition, this line slants up to the right, so the slope of the line is positive. This checks out, since the slope of the line is $+\frac{1}{3}$ and not $-\frac{1}{3}$.

The slope (m) of the line is and the y-intercept (b) is 1. Thus, the equation is $y = \frac{1}{3}x + 1$, which is choice (E).

Answer choice (A) has a slope of $\frac{1}{3}$, but a y-intercept of –1 which is incorrect. Choice (B) has a slope of $-\frac{1}{3}$ and a y-intercept of –1, which are both incorrect. Choice (C) has a y-intercept of –1 and a slope of 3, so each of these values is also incorrect. Choice (D) has a slope of –3 and a y-intercept of 1. Even though the y-intercept is correct, this line would slant up to the left and be too steep to be the line in the question.

22. B

Chapter 9 describes the various types of equations that you may encounter on the coordinate plane. One of these is the circle. The coordinates of the center of the circle and the radius determine the equation of a circle. In this graph, the center of the circle is at (5, 4). Count the number of spaces that make up the radius to conclude that the radius of this circle is 4 units. The equation of a circle has the form $(x - x_1)^2 + (y - y_1)^2 = r^2$, where (x_1, y_1) is the center of the circle and r is the radius. The correct equation is $(x - 5)^2 + (y - 4)^2 = 4^2$, or $(x - 5)^2 + (y - 4)^2 = 16$.

If you answered (A), you forgot to square the radius in the equation. This is a common error. If your choice was (C), you did not square the radius, and be very careful—notice that only the variables x and y are squared; the entire expression in parentheses should be squared. If you were working too quickly, you might easily have chosen choice (D). But again notice that, just like with choice (C), only the variables x and y are squared. In choice (E), the operators in the parentheses are plus signs. If this were correct, the center of the circle would have been at (–5,–4).

23. C

This equation is a quadratic equation; quadratic equations are of the form $ax^2 + bx + c$, where *a, b,* and *c* are coefficients (constants). A quadratic equation forms a figure on the coordinate plane called a *parabola*. This concept is covered in Chapter 9. A parabola has a ⋃ or ⋂ shape; it is not a straight line like a linear equation. The *vertex* of the graph of a quadratic equation is also called the *turning point*, or the *maximum* or *minimum point*. If the shape is a ⋃, the parabola opens upwards, and the vertex is the bottommost point (the minimum). If the shape is ⋂, the parabola opens downwards, and the vertex is the topmost point (the maximum).

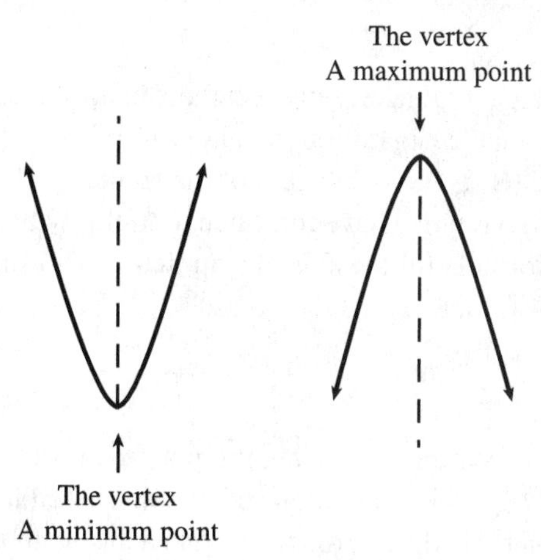

The vertex lies on the axis of symmetry, the straight line that cuts the figure in half, as shown on the graphs above as a dotted line down the middle of each parabola. To find the coordinates of the vertex point of a quadratic equation, first find the equation of the axis of symmetry. This equation is found by using the coefficients *a* and *b*.

The equation is $x = -\frac{b}{2a}$, or $x = -\frac{-12}{2 \bullet 3}$. This simplifies to $x = 2$. The x-coordinate of the vertex is 2, because it lies on this axis of symmetry. To find the y-coordinate, substitute in the value of 2 into the given equation $y = 3x^2 - 12x + 6$, which is $y = 3 \bullet 2^2 - 12 \bullet 2 + 6 = 12 - 24 + 6 = -12 + 6 = -6$. The coordinates of the vertex are (2,–6). The figure below shows the graph of the equation $y = 3x^2 - 12x + 6$.

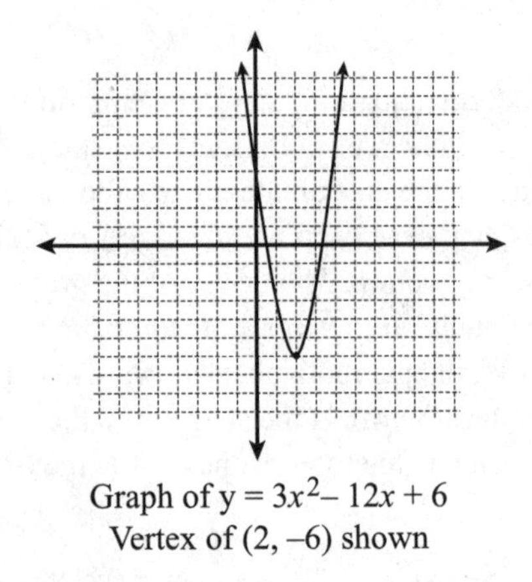

Graph of $y = 3x^2 - 12x + 6$
Vertex of (2, –6) shown

If you chose choice (A), you incorrectly had the x-coordinate to be –2; you may have forgotten the negative sign in the formula to find the axis of symmetry. Choice (B) is the y-intercept of the equation. If your answer was choice (D) or (E), you may have forgotten to multiply by two in the denominator of the formula for the axis of symmetry, resulting in an x-coordinate of 4 or –4 for the vertex.

24. C

The number of lines of symmetry in a regular polygon is determined by the number of sides and angles it contains, a concept discussed in Chapter 10. When a *line of symmetry* is drawn, the figure on each side of the line is a congruent "mirror image" of the other side.

A regular pentagon is a five-sided polygon whose sides are all congruent, or all the same length, and whose angles are all the same measures. A regular pentagon has five lines of symmetry as shown in the figure below. Each line of symmetry bisects an angle and the side opposite that angle in the pentagon.

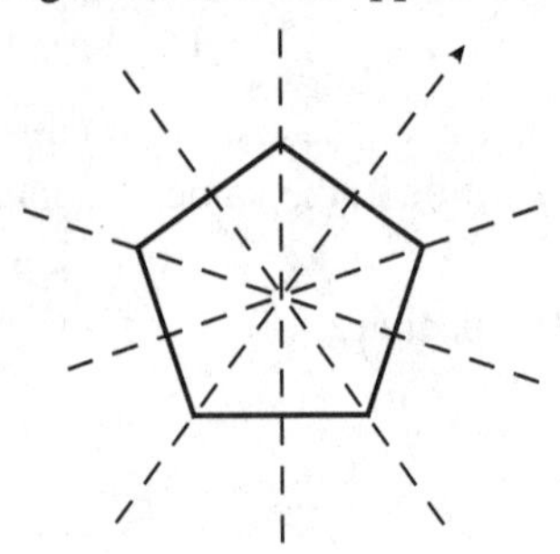

25. C

There are four basic types of transformations: reflections, rotations, translations, and dilations. Each is explained in detail in Chapter 10. A *reflection* gives a mirror image of an object. A *rotation* takes an object and turns it a certain number of degrees. A *translation* slides an object a certain number of units in a certain direction. A *dilation* enlarges or shrinks an object based on a scale factor.

In this question, the triangle is enlarged as each side of triangle ABC is twice the size of the sides of triangle ABC. The only type of transformation that changes the size of the original figure is a dilation. Since the image is twice as large, the correct transformation is a dilation.

26. D

The *locus of points* is the set of points that fit a certain criteria. This concept is explained and examples are given in Chapter 11. When two or more conditions of a locus of points are used within the same problem, this is known as *compound loci.* This question asks for the number of points that fit two sets of criteria, so each set will need to be examined separately first. Then, the intersection, or points that satisfy both conditions, can be determined for the compound loci.

The first step in this question is to find the locus of points 3 units from line *l*. The locus is formed by two lines that run on either side of line *l*, each parallel to line *l*, and each 3 units away. The locus could appear as the figure below:

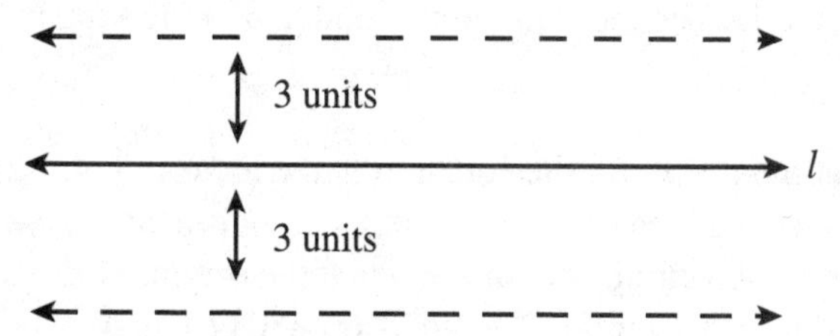

Next, take a look at the other condition. The locus of points 5 units from a point P on line *l* will be a circle that surrounds point P with a radius of 5. This would appear as the figure below:

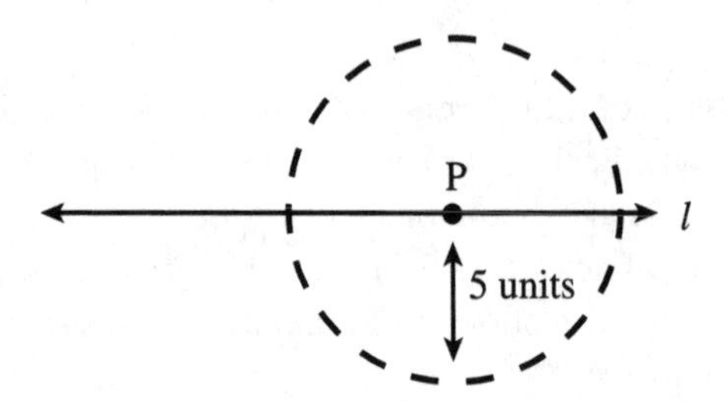

Since the question asked for the number of points in the intersection, put the two conditions together and see how many places the locus of points overlap. The compound loci from this problem are pictured together below:

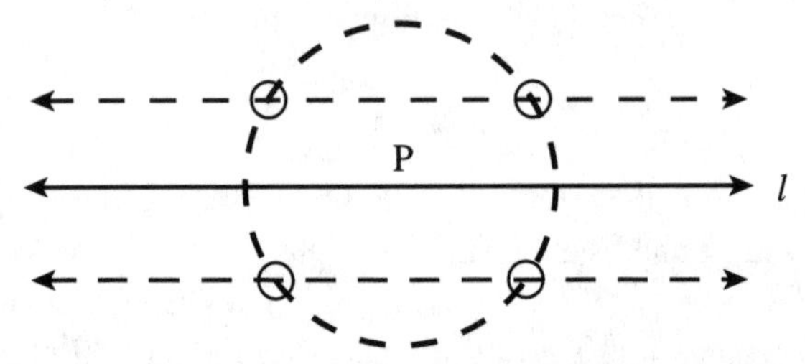

As seen in the figure, the dotted lines intersect in four places shown by the four circles. Therefore, there are four points that satisfy the compound loci in this question.

27. B

The *locus of points* about a point is a circle that surrounds the point. Each point is equidistant from the given point. This given point becomes the center of the circle. The locus of points about a point is explained in detail in Chapter 11.

For this question, it asks for the equation of the locus of points 6 units from the origin. Since the locus of points about a point is a circle, use the form for the equation of a circle with the center at the origin. This is $x^2 + y^2 = r^2$, where r = radius of the circle. The radius (r) will be equal to the number of units the locus is away from the point. Therefore, r = 6 units in this question.

Substitute $r = 6$ into the equation $x^2 + y^2 = r^2$: $x^2 + y^2 = 36$

Choice (A) is the result of substituting $r = 6$, but forgetting to square the 6 to make 36. Choices (C), (D), and (E) are not correct because in order to be an equation of a circle, both the x and y must be squared in the equation of a circle.

28. E

This figure is a right triangle, and right triangle geometry is covered in Chapter 12. Miguel wants to enclose a triangular area with fencing, so he will need an amount of fencing equal to the perimeter of the triangle. The measures of two of the sides of the triangle are given, and you need to determine the length of the third side. Because the triangle is a right triangle, use the Pythagorean theorem to find the missing side. In this triangle, the sides measuring 10 and 12 are the legs of the right triangle; the missing measurement is the length of the hypotenuse. The *Pythagorean theorem* states that the sum of the squares of the legs in a right triangle is equal to the square of the hypotenuse. This is often written as $a^2 + b^2 = c^2$, where a and b are the legs of a right triangle and c is the hypotenuse.

Substitute in the known values: $10^2 + 12^2 = c^2$, or $100 + 144 = c^2$.
Add on the left hand side: $244 = c^2$
Take the square root of both sides of the equation: $c = \sqrt{244} = 15.62$, rounded to the nearest hundredth.

This is the length of the hypotenuse. Now, finish the problem by finding the perimeter. *Perimeter*, covered in Chapter 5, is the distance around a figure. In this case, it is $10 + 12 + 15.62 = 37.62$ ft.

If your answer was choice (A) or choice (D), you used the Pythagorean theorem but mistakenly thought that the length of 12 feet represented the hypotenuse. Choice (A) would be the incorrect length, and choice (D) would be the incorrect perimeter. If you chose (C), you simply added the two given lengths. Choice (B) is the correct length for the hypotenuse, but it is not the total amount of fencing needed.

29. B

When you are given a figure that is a right triangle, with one angle measure and one side measure given, you will most often need to use trigonometry to solve the problem. Trigonometry is covered in Chapter 12. In this right triangle, side NL is the hypotenuse of the triangle. Side NM is the side that is adjacent to the angle measure marked 30°. Trigonometric ratios are ratios that compare sides of the right triangle. The ratio you will use to find NM will be the *cosine ratio,* which is $\frac{adjacent}{hypotenuse}$. Let x represent the length of NM, the side adjacent to the angle of 30°.

Set up the ratio as: $\cos 30° = \frac{x}{24}$.

Isolate the variable x by multiplying both sides of this equation by 24.

The equation becomes: $24(\cos 30°) = x$

Use your calculator to multiply: $x = 20.78$ m

If you chose (C) you gave the length of NL, not NM. Choice (A) is the length of side ML, and you may have incorrectly solved for this side, by using the sine ratio. If your choice was (E), you most likely forgot about trigonometry altogether; in this case, be sure to study Chapter 12.

30. B

This is a figure of a right triangle. Two of the side lengths are listed, and you are asked to find the measure of an angle of the triangle. When problems involve a right triangle and an angle in this triangle, you should immediately think of trigonometry, which is described in Chapter 12. In relation to ∠DEC, side DC is opposite to this angle, and side EC is the hypotenuse of the right triangle. *Trigonometric ratios* are ratios that compare sides of the right triangle. The ratio you will use to find NM will be the sine ratio, which is $\frac{opposite}{hypotenuse}$. Let x represent the angle measure.

Set up the ratio as $\sin x = \frac{8}{16}$, or $\sin x = 0.5$.

To find the angle measure, use the arcsin function on your calculator, often shown as the $\sin^{-1}$ key. So $x = \arcsin(0.5)$, and $x = 30°$.

If your choice was (C), you may have thought that this was a right isosceles triangle, described in Chapter 2. Choice (D) is the measure of angle∠EDC, the right angle of the triangle. If your choice was (E), you most likely forgot about trigonometry altogether; in this case, be sure to study Chapter 12.

CHAPTER 1

Angles and Special Angle Pairs

WHAT ARE ANGLES AND SPECIAL ANGLE PAIRS?

There are many basic figures of geometry you need to understand before studying more complex concepts. This chapter will introduce those basic figures and how they are used to construct just about every other geometric figure. Angles can be found everywhere in our surroundings and provide a good stepping stone to begin your study of geometry. They appear in every drawing, every piece of furniture, and many elements of nature; just about anywhere you look you see angles. This chapter will define the different types of angles and also explore some of the special angle-pair relationships that make working with angles a little easier.

Chapter 1 will include basic concepts of geometry, as well as important steps to remember as you work through geometry problems involving angles. The five most common angle and angle-pair questions you may encounter will also be explained in detail to help you on your way. As you will soon see, these basic figures and concepts are essential elements of geometry and will be embedded in each of the other chapters in this book.

CONCEPTS TO HELP YOU

This chapter will start with a review of the building blocks of geometry, including the main topic of angles. Classifying, or naming, angles will be covered as well as special angle pairs. These pairs are formed when two or more lines intersect or when parallel lines are cut by another line. Recognizing the patterns that emerge with these special angle pairs can be very useful when solving problems involving lines and angles.

Basic Terms of Geometry

Let's begin our study of geometry with some of the basic figures: point, line, and plane. These are known as the *building blocks of geometry*.

A *point* is a location on a plane or in space; a point has no thickness. A point is named by one letter, as in point A in the figure below.

• A

A *line* is the set of all points in a plane in a single path that continues in opposite directions; a line has no thickness or width. A line is usually named by using two points on the line. Some of the ways the line below can be named are $\overline{AB}$, $\overline{BA}$, $\overline{BC}$, $\overline{CB}$, $\overline{AC}$ or $\overline{CA}$.

A B C

A *plane* is the set of all points on a flat surface that extends in all directions; a plane has no thickness. A plane is named using three noncollinear points located in the plane, or three points not contained on the same line. Plane ABC is shown in the figure below.

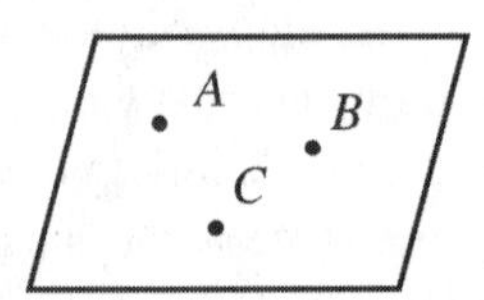

These three terms help us to define the rest of the figures of geometry.

Let's move on to some of the figures these building blocks help us define. The next term is *line segment.* A line segment is part of a line. Any line segment is named by its two endpoints. Line segments $\overline{AB}$ and $\overline{CD}$ are shown in the figure below. Line segments make up the sides and edges of many geometric figures, such as polygons and solids that will be discussed in later chapters.

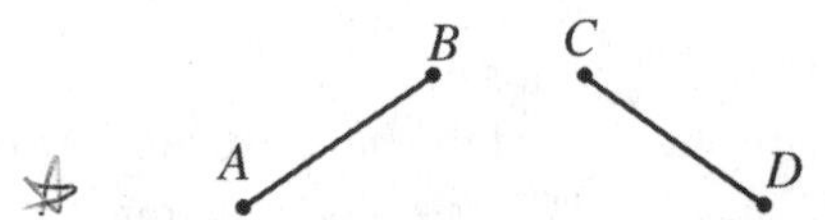

The next term is *ray*. Since a line is defined as the set of points that follow a single path in opposite directions, a ray can be described as half of a line. A ray is the set of all points in a single path beginning at an endpoint and continuing in one direction. Rays are named by their endpoint and at least one other point on the ray. For example, the ray in the figure below can be named as ray $\overrightarrow{AB}$, $\overrightarrow{AC}$, or $\overrightarrow{AD}$.

A B C D

Having defined these basic terms, we come to the major focus of this chapter, angles. An *angle* is the joining of two rays at a common endpoint.

This endpoint is known as the *vertex* of the angle. Angles can be named by using just the vertex of an angle, a point on each of the rays of the angle and the vertex, or just a number in the interior of the angle. When using three letters to name the angle, the order of the letters should always list the vertex as the middle letter. The figure below shows a few examples of how angles can be named.

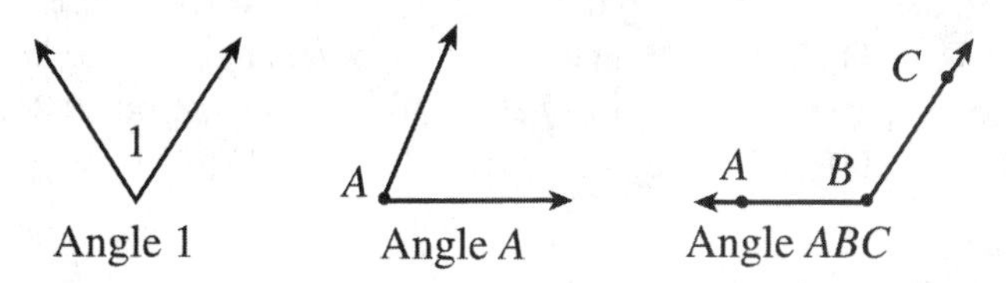

Classifying Angles

Angles can be classified, or named, based on the number of degrees in the angle. There are five common angle types that are defined below:

1. An acute angle has a measure between 0 and 90 degrees.
2. A right angle has a measure of 90 degrees.
3. An obtuse angle has a measure between 90 and 180 degrees.
4. A straight angle has a measure of 180 degrees.
5. A reflex angle has a measure greater than 180 degrees.

An example of each type is shown in the figure below.

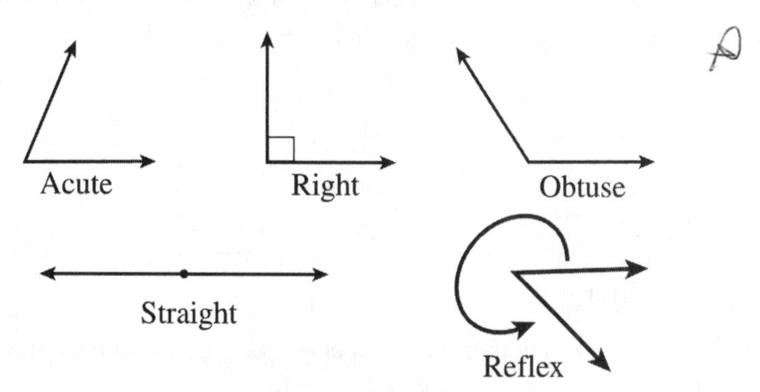

Some of these angle types are used to classify triangles, as you will see in Chapter 2.

SPECIAL ANGLE PAIRS

Complementary Angles 

One of the special types of angle pairs is *complementary angles*. The sum of two complementary angles is 90 degrees. The angles do not have to be adjacent, or next to each other, to be complementary, but if placed next to each other, they will form a right angle. If two angles are complementary, one angle is said to be the complement of the other angle. Some examples of complementary pairs, as shown below, are 40 degrees and 50 degrees, since 40 + 50 = 90, and 15 degrees and 75 degrees, since 15 + 75 = 90.

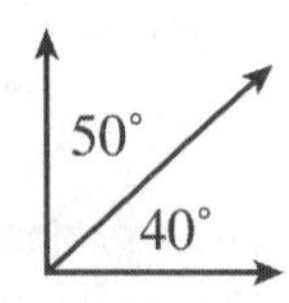

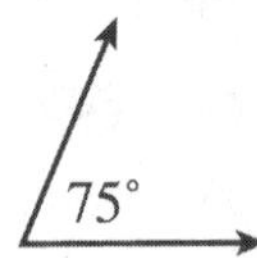

Supplementary Angles

Another special type of angle pairs is *supplementary angles*. The sum of two supplementary angles is 180 degrees. These angles also do not have to be adjacent to be supplementary but, if placed next to each other, will form a straight angle. If the two supplementary angles are adjacent to each other, they are called a *linear pair*. If two angles are supplementary, one angle is said to be the *supplement* of the other angle. Some examples of supplementary pairs, as shown below, are 30 degrees and 150 degrees, since 30 + 150 = 180, and 90 degrees and 90 degrees, since 90 + 90 = 180. The first example is a linear pair since the angles are also adjacent to each other and share a side.

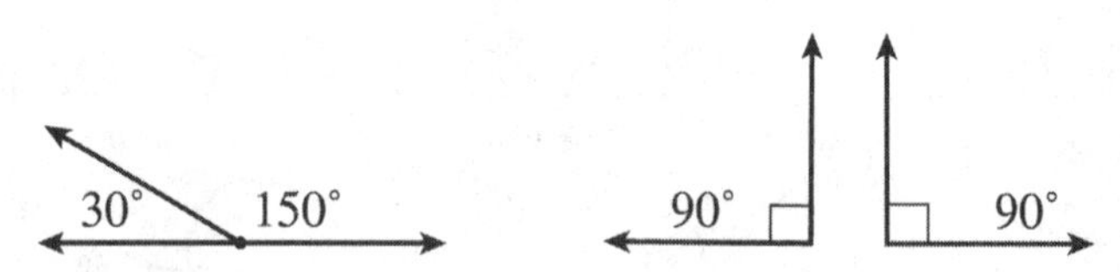

Vertical Angles

Vertical angles are another type of special angle pair. They are the nonadjacent angles that are formed when two lines intersect. Instead of having a sum measure of 90 or 180 degrees, vertical angles always have the same measure. The figure below shows two pairs of vertical angles. Lines l and m intersect to form angles 1, 2, 3, and 4. Angles 1 and 3 are vertical angles and angles 2 and 4 are vertical angles. Note the patterns in the measures of the vertical angles. Angles 1 and 3 are both 60 degrees and angles 2 and 4 are both 120 degrees.

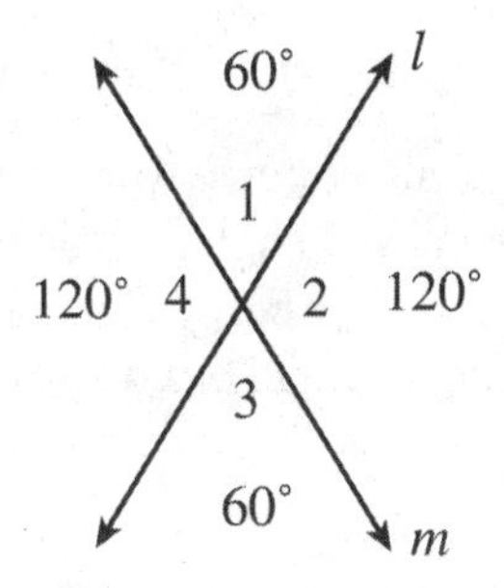

SPECIAL ANGLE PAIRS FORMED BY PARALLEL LINES

Congruent Angle Pairs with Parallel Lines

Before discussing the angles formed by parallel lines, let's first define the term *parallel lines*. Parallel lines are two or more lines in the same plane that will never intersect. People often think of railroad tracks or the double yellow lines found on most roads as examples of parallel lines found in everyday life. There are important angle relationships that develop when two parallel lines are cut, or crossed, by another line that is not parallel to them. These special relationships are corresponding angles, alternate interior angles, alternate exterior angles, and as mentioned before, vertical angles. Use the figure below to help visualize these different cases of angle pairs.

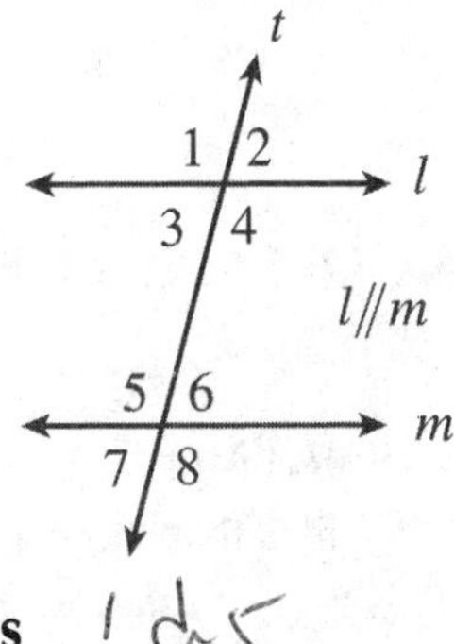

Corresponding Angles

Corresponding angles are formed when a transversal cuts two or more other lines. Each line forms a corresponding angle with the transversal. The corresponding angles are on the same side of the transversal on the same side of the parallel line that forms each. In the figure above, angles 1 and 5,

2 and 6, 3 and 7, and 4 and 8 are corresponding angle pairs. If the lines are parallel, then the corresponding angles are congruent.

Alternate Interior Angles

Alternate interior angles are also formed when a transversal cuts two or more other lines. Alternate interior angles are on opposite sides of the transversal but on the interior of, or between, the parallel lines. In the figure with parallel lines above, angles 3 and 6 and angles 4 and 5 are alternate interior angles. If the lines are parallel, then the alternate interior angles are congruent.

Alternate Exterior Angles

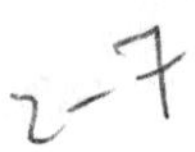

Alternate exterior angles are also formed when a transversal cuts two or more other lines. Alternate exterior angles are on opposite sides of the transversal but on the exterior, or outside, of the parallel lines. In the figure with parallel lines above, angles 1 and 8 and angles 2 and 7 are alternate exterior angles. If the lines are parallel, then the alternate exterior angles are congruent.

Vertical Angles

It is worth mentioning here the patterns of vertical angles when parallel lines are cut by a transversal. Recall that vertical angles are the nonadjacent angles formed by two intersecting lines. Vertical angles are congruent. In the figure above, angles 1 and 4, 2 and 3, 5 and 8, and 6 and 7 are all vertical angle pairs.

SUPPLEMENTARY ANGLE PAIRS FORMED BY PARALLEL LINES

Adjacent Angles and Linear Pairs

When two adjacent angles, or angles that share a side, are supplementary, they form a straight angle. This type of relationship is called a *linear pair.* Any pair of adjacent angles in the above diagram forms a linear pair and has a sum of 180 degrees. Angles 1 and 2, 2 and 4, 1 and 3, 3 and 4, 5 and 6, 6 and 8, 5 and 7, and 7 and 8 are the eight linear pairs in the above diagram.

Interior Angles on the Same Side of the Transversal

Interior angles on the same side of the transversal, but between the two parallel lines, will be supplementary. Angles 3 and 5 and angles 4 and 6 from the figure above are interior angles on the same side of the transversal. The sum of their measures is equal to 180 degrees.

Exterior Angles on the Same Side of the Transversal

As with the interior angles on the same side of the transversal, exterior angles on the same side of the transversal will also be supplementary. Angles 1 and 7 and angles 2 and 8 from the figure above are exterior angles on the same side of the transversal. The sum of their measures is equal to 180 degrees.

A Special Case: Perpendicular Lines

An additional special case for lines cut by a transversal involves perpendicular lines. *Perpendicular lines* are lines that meet to form right angles. Therefore, if two parallel lines are crossed by a perpendicular transversal, all angles, no matter what the relationship, would be equal to 90 degrees. This is shown in the figure below.

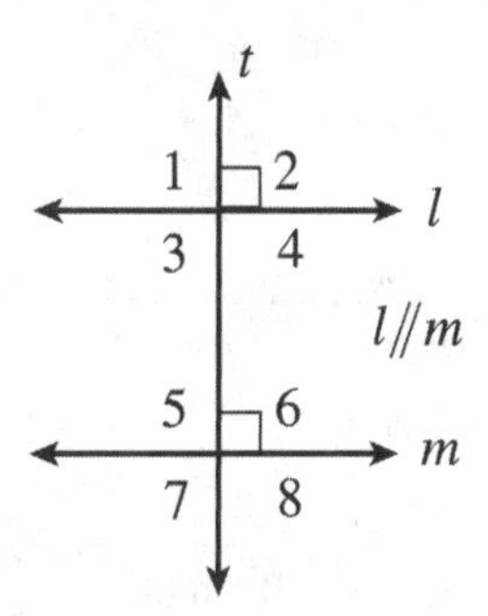

STEPS YOU NEED TO REMEMBER

Using Ratios to Solve Angle Problems

A common word problem involving angle pairs gives clues about the angles with ratios. For example, a problem may have you solve for the smaller angle of two complementary angles whose ratio is 2:7. One strategy to use when solving a question such as this is to take each number of the ratio and multiply it by x to get an equivalent ratio. Therefore, the ratio of 2:7 would become $2x$ and $7x$. Now the two numbers can be added in an equation and set equal to 90 since the sum of two complementary angles is 90 degrees.

The equation would become: $2x + 7x = 90$
Combine like terms to get: $9x = 90$
Divide each side of the equation by 9 to get: $\frac{9x}{9} = \frac{90}{9}$
The value of x is: $x = 10$
Since the smaller angle was represented by $2x$,
multiply the value of x by 2: $2x = 2(10)$
The smaller angle is 20 degrees: $x = 20$

Using this strategy in other situations with ratios will also be discussed in later chapters.

Patterns with Parallel Line Diagrams

There are a number of patterns that occur in parallel line diagrams that can make solving for angles within the diagram easier. One major pattern deals with the fact that when the transversal crosses the parallel lines, eight angles are created. Of these eight angles, four will be acute angles and four will be obtuse. These angles alternate in the figure, as shown below. Each of the acute angles will be congruent in measure and each of the obtuse angles will be congruent in measure.

t
obtuse acute
l
acute obtuse
m//l
obtuse acute
m
acute obtuse

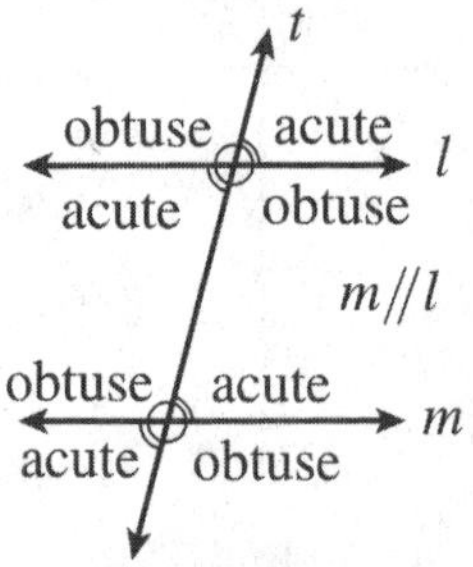

The only case where this would not be true is if the transversal is perpendicular to the parallel lines, and all eight angles would then be equal to 90 degrees.

Using Algebra by Setting up Equations to Solve Angle-Pair Problems

Equation solving is an important strategy to use when solving many angle-pair problems. This section will take you through how to set up equations to solve this type of question. Solving these equations will be explained in detail in the next section.

The task of setting up an equation can seem a bit daunting at first, but can be made less intimidating if you realize that most angle-pair problems boil down to one of three scenarios:

1. The two angles in the relationship are equal in measure.
2. The two angles in the relationship are supplementary.
3. The two angles in the relationship are complementary.

In case 1, you may have a pair of corresponding angles, alternate interior angles, alternate exterior angles, or vertical angles. In these instances, take the number or expressions for each and set them equal to each other. Then solve for the variable, if necessary. For example, if two alternate interior angles are labeled $3x - 2$ and $6x - 23$, the equation would be:

$$3x - 2 = 6x - 23.$$

In case 2, you have a pair of supplementary angles. This time, take the expressions for each and add them, setting the equation equal to 180 degrees. For example, if two supplementary angles are represented by $4x + 5$ and $8x - 10$, the equation would be:

$$4x + 5 + 8x - 10 = 180.$$

In case 3, you have a pair of complementary angles, which have a sum of 90 degrees. Take the sum of the expressions of the angle measures and set it equal to 90. For example, if the complementary angles labeled $2x$ and $9x - 24$, the equation would be:

$$2x + 9x - 24 = 90.$$

Be careful in questions such as these to answer the question being asked. Sometimes you are asked just to solve for x; other times you are asked for the measure of one or both of the angles. In this case, you would need to substitute the value of x into one or both of the expressions to find the angle measure. Always read the question again before answering to make sure you have found what you were asked to look for!

STEP-BY-STEP ILLUSTRATION OF THE FIVE MOST COMMON QUESTION TYPES

This next section will give you examples of the some of the most common question types involving angles and special angle pairs you may encounter. Remember, these questions will set the stage for some of the more advanced concepts you see in the chapters that follow.

Question 1: Solving Complementary/Supplementary Angle Pairs with Algebra

Two angles are complementary. If the measure of one angle is 20 degrees more than the measure of its complement, what is the measure of the larger angle?

(A) 35°

(B) 55°

(C) 70°

(D) 80°

(E) 90°

This question involves complementary angles, a concept that was explained earlier in the chapter. Two complementary angles have a sum of 90 degrees. In this question, you are not given any of the angle measures. Instead, you are given clues as to how the two angles are related. Using algebra by writing an equation is an effective way to approach this question.

Start by using x to represent the number of degrees in the smaller angle. Since one angle is 20 degrees more than the other angle, then the larger angle can be represented by $x + 20$. The setup for this type of question was demonstrated in the previous section. Since the two angles are complementary, add the expressions used for each angle and set the sum equal to 90.

The equation could be set up as:	$x + x + 20 = 90$
Combine like terms in the equation:	$2x + 20 = 90$
Subtract 20 from each side of the equation:	$2x + 20 - 20 = 90 - 20$
The equation becomes:	$2x = 70$
Divide each side of the equation by 2 to get the x alone:	$x = 35$

The measure of x, which represents the smaller angle, is 35 degrees. This question asked for the measure of the larger angle, which is represented by $x + 20$. Thus, the measure of the larger angle is $35 + 20 = 55$ degrees.

The correct answer is choice (B). Choice (A) is the number of degrees in the smaller angle, but the question asked for the measure of the larger angle. Choice (C) is the value in the equation that was set equal to $2x$. If you forgot to divide each side of the equation by 2, you may have selected this choice. Choice (D) is the value of x if the equation was mistakenly set equal to 180 degrees, instead of 90. If this question described the two angles as supplementary, this would have been the measure of the smaller angle. Choice (E) is the sum of the two angles, but not the measure of the larger angle.

Question 2: Solving Complementary/Supplementary Angle Pairs with Ratios

Two supplementary angles are in the ratio 1:5. What is the measure of the smaller angle?

(A) 15°

(B) 30°

(C) 75°

(D) 150°

(E) 180°

This question involves supplementary angles. This type of angle pair has a sum of 180 degrees. The clues for how the pair is related are given in ratio form. The angles are in the ratio 1:5. As explained in an earlier section, one strategy to use with ratios is to multiply each number in the ratio by x. The expressions in the ratio would then become $1x$ and $5x$, where the smaller number is represented by $1x$ and the larger number is represented by $5x$.

Next, write an equation using the expressions for the two angles. Since the two angles are supplementary, add the two expressions and set the sum equal to 180 degrees.

The equation could be:	$1x + 5x = 180$
Combine like terms in the equation:	$6x = 180$
Divide each side of the equation by 6 to get the x alone:	$x = 30$

Since the measure of x is 30, and x represents the smaller angle, the measure of the smaller angle is 30 degrees. **The correct answer is choice (B).**

You may have selected choice (A) if the equation was set equal to 90 degrees, instead of 180 degrees. This would have been the correct answer if the pair of angles was described as complementary, instead of supplementary. Choice (C) is the measure of the complement of a 15-degree angle. Choice (D) is the measure of the larger angle of this supplementary pair; however, this question was looking for the smaller angle. Choice (E) is the sum of the two supplementary angles.

Question 3: Vertical Angle Problems with Algebra

In the figure below, line *l* intersects with line *m*. What is the value of angle $\angle ABC$?

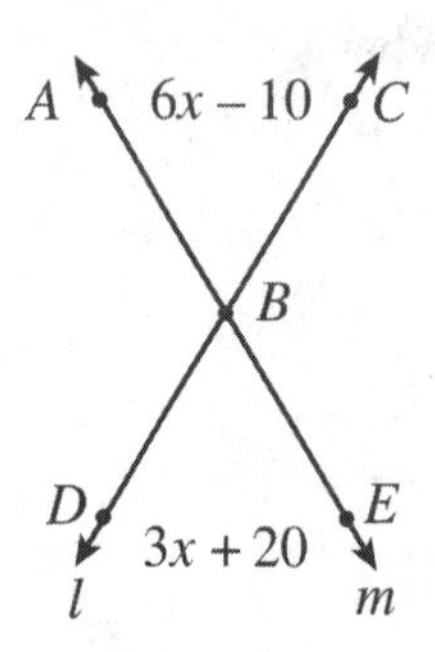

(A) 10°

(B) 18.8°

(C) 40°

(D) 50°

(E) 130°

When any two lines intersect, four angles are formed. The two nonadjacent angles, or the angles that are not next to each other, are called *vertical angles*. Vertical angles are congruent, or have the same measure. In this question, you are given information about the angles in expression form. One of the angles is represented by the expression $6x - 10$ and the other expression is represented by $3x + 20$. These expressions can be used with algebra in an equation in order to find the value of x.

Since the measures of the vertical angles are equal, take the two expressions for the angle and set them equal to each other in an equation. Then, solve the equation to find the value of x in order to find the correct answer choice.

The equation can be written as:	$6x - 10 = 3x + 20$
Subtract $3x$ from each side of the equation:	$6x - 3x - 10 = 3x - 3x + 20$
The equation simplifies to:	$3x - 10 = 20$
Add 10 to each side of the equation:	$3x - 10 + 10 = 20 + 10$
The equation becomes:	$3x = 30$
Divide each side of the equal sign by 3 to get x alone:	$\frac{3x}{3} = \frac{30}{3}$
	$x = 10$

This question asks for the measure of angle $\angle ABC$. This angle is represented by the expression $6x - 10$.

Substitute $x = 10$ into this expression to find the number of degrees in the angle:	$6x - 10$
Multiply first, then subtract:	$6(10) - 10$
	$60 - 10 = 50$

The number of degrees in angle $\angle ABC$ is 50 degrees. **The correct choice is (D).** Choice (A) is the correct value of x, but the question asks for the number of degrees in angle $\angle ABC$. Choice (B) is the result of finding the sum of the expressions and setting it equal to 180 degrees. Choice (E) is the supplement of a 50-degree angle and would represent the two other angles in the diagram.

Question 4: Solving for Angle Measures with Parallel Lines—Congruent Angles

In the figure below, parallel lines *m* and *n* are crossed by transversal *t*. What is the measure of angle ∠CFG?

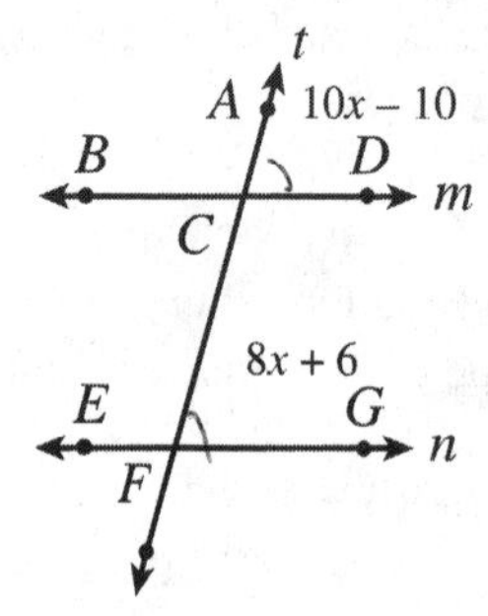

(A) 8°

(B) 20°

(C) 70°

(D) 110°

(E) 140°

There are many patterns that emerge in a figure where two parallel lines are cut, or crossed, by another line. In these figures, eight angles are formed. Unless the transversal is perpendicular, four of these angles are acute in measure and all four are congruent to each other and the other four are obtuse in measure and are also congruent to each other.

In addition, one of the acute angles paired with one of the obtuse angles forms a supplementary pair. The first step, therefore, is to determine if the angles given in the diagram are congruent or supplementary to each other.

In this figure, the two given angles are corresponding. They are on the same side of the transversal and each are above the parallel line that forms them. Since they are corresponding and the lines are parallel, their measures are equal.

To find the measure of angle ∠CFG, first solve for *x*. Use algebra and write an equation to show the relationship between the two given angles in the figure. The measures of the two angles are equal, so set the two expressions equal to each other in the equation.

The equation could be set up as:	$10x - 10 = 8x + 6$
Subtract $8x$ from each side of the equation:	$10x - 8x - 10 = 8x - 8x + 6$
The equation becomes:	$2x - 10 = 6$
Add 10 to each side of the equation:	$2x - 10 + 10 = 6 + 10$
The equation is now:	$2x = 16$
Divide each side by 2 to get x alone:	$\frac{2x}{2} = \frac{16}{2}$
	$x = 8$

Since the value of x is 8, substitute this value into the expression that represents the measure of angle ∠CFG. This expression is $8x + 6$. Therefore, the measure of the angle is: $8(8) + 6 = 64 + 6 = 70$ degrees.

Choice (C) is the correct answer. Choice (A) is the correct value of x, but not the measure of the angle stated in the question. Choice (B) is the complement of a 70-degree angle. Choice (D) is the measure of each of the obtuse angles in the figure, but this question was looking for an acute angle. Choice (E) is the sum of the measures of the two angles labeled in the figure.

Question 5: Solving for Angle Measures with Parallel Lines—Supplementary Pairs

In the figure below, parallel lines r and s are cut by transversal t. The measure of angle ∠ACD is represented by $7x$ and the measure of angle ∠CFE is represented by $3x + 40$. What is the measure of the smaller angle?

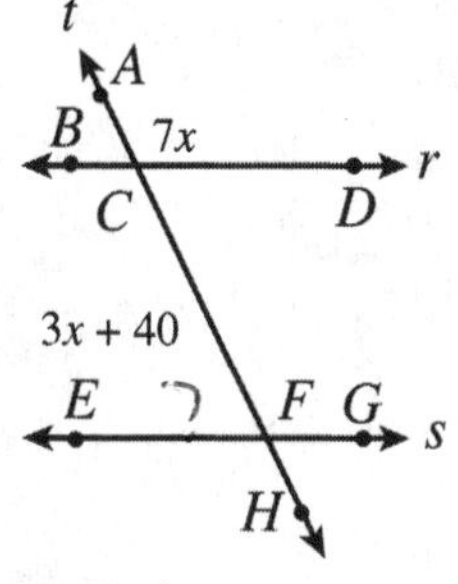

(A) 10°

(B) 14°

(C) 70°

(D) 82°

(E) 98°

As stated in question 4, there are many patterns that emerge in a figure where two parallel lines are cut by another line. In these figures, eight angles are formed. Unless the transversal is perpendicular, four of these angles are acute in measure and all four are congruent to each other and the other four are obtuse in measure and are also congruent to each other. In addition, one of the acute angles paired with one of the obtuse angles forms a supplementary pair. The first step in this type of question is to determine if the angles given in the diagram are congruent or supplementary to each other.

In this figure, the two given angles are supplementary. One way to draw this conclusion is to use the special angle pairs studied in this chapter. One of these special angle pairs was vertical angles. In this diagram, angle $\angle ACD$ is vertical to angle $\angle BCF$. Therefore, these two angles are congruent. Now, examine the angle pair $\angle BCF$ and $\angle CFE$. Each of these angles is on the same side of the transversal, and each is on the interior of the parallel lines. These two angles are supplementary. The sum of their measures is 180 degrees.

To find the measure of the angles, first solve for x. Then substitute this value into each expression to find the measure of each angle. Keep in mind, this question is looking for the measure of the smaller angle.

Since the angles are supplementary, add the expressions and set the sum equal to 180 degrees.

The equation could be written as:	$7x + 3x + 40 = 180$
Combine like terms:	$10x + 40 = 180$
Subtract 40 from each side of the equation:	$10x + 40 - 40 = 180 - 40$
The equation becomes:	$10x = 140$
Divide each side by 10 to get the x alone:	$\frac{10x}{10} = \frac{140}{10}$
	$x = 14$

Since the value of x is 14, substitute into each expression to find the measure of each angle.

The measure of angle $\angle ACD = 7x = 7(14) = 98$ degrees. The measure of angle $\angle CFE = 3x + 40 = 3(14) + 40 = 42 + 40 = 82$ degrees. The measure of the smaller angle is 82 degrees, so **the correct answer is choice (D).**

Choice (A) is the value of x if the two expressions in the figure were set equal to each other. Since the two angles are supplementary and not congruent, this is not the correct answer. Choice (C) is the result of using $x = 10$ and substituting into either expression for the two angles. Choice (B) is the correct value of x when the equation is solved, but not the measure of the smaller angle. Choice (E) is the measure of the larger of the two angles, but the question asked for the smaller angle.

CHAPTER QUIZ

1. Which of the following best describes an angle that measures 123°?

   (A) Acute

   (B) Reflex

   (C) Obtuse

   (D) Right

   (E) Straight

2. Two angles are complementary. If the measure of one angle is 65°, what is the measure of the other angle?

   (A) 25°

   (B) 65°

   (C) 90°

   (D) 115°

   (E) 180°

3. What is the value of x in the figure below?

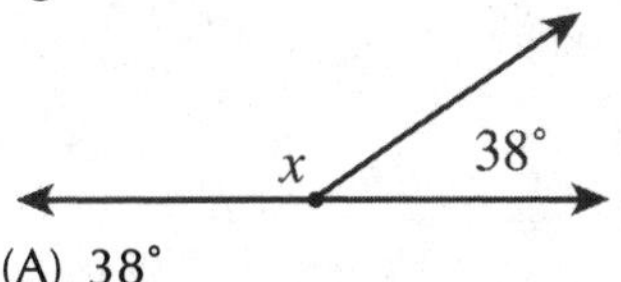

   (A) 38°

   (B) 52°

   (C) 90°

   (D) 142°

   (E) 152°

4. Two complementary angles are in the ratio 2:7. What is the measure of the larger angle?

   (A) 10°

   (B) 20°

   (C) 40°

   (D) 70°

   (E) 140°

5. The measure of one angle is 30 degrees more than twice its supplement. What is the measure of the smaller angle?

   (A) 30°

   (B) 50°

   (C) 75°

   (D) 100°

   (E) 130°

6. What is the measure of angle 1 in the figure below?

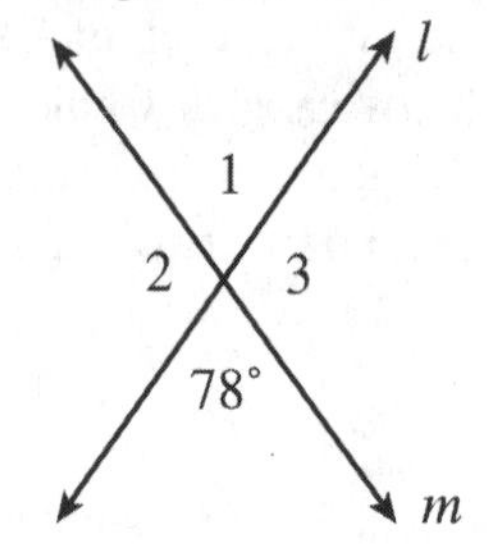

(A) 12°
(B) 78°
(C) 79°
(D) 102°
(E) 282°

7. Lines *r* and *s* intersect in the figure below. What is the measure of angle ∠ABD?

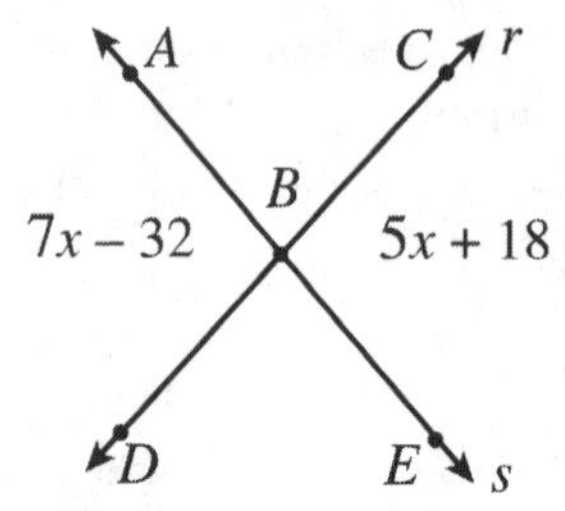

(A) 16.16°
(B) 25°
(C) 37°
(D) 118°
(E) 143°

8. Two alternate interior angles formed by the same two parallel lines and transversal are expressed as $8x - 2$ and $3x + 17$. What is the measure of the angles?
(A) 4°
(B) 15.1°
(C) 29°
(D) 61°
(E) 62.3°

9. In the figure below, two parallel lines *l* and *m* are cut by transversal *n*. What is the measure of angle 1?

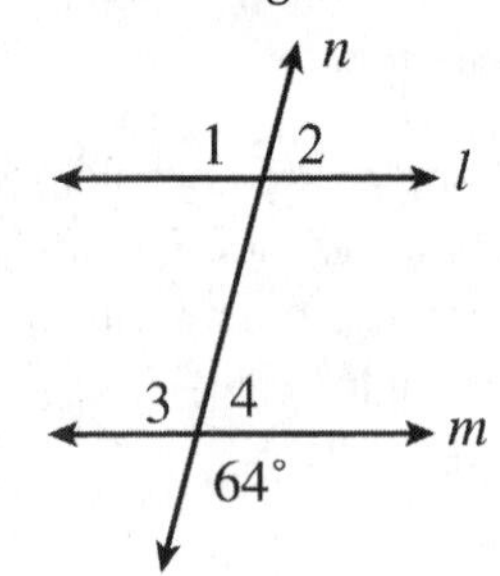

(A) 26°
(B) 64°
(C) 116°
(D) 126°
(E) 180°

10. In the figure below, two parallel lines *l* and *m* are cut by transversal *t*. What is the measure of the larger angle?

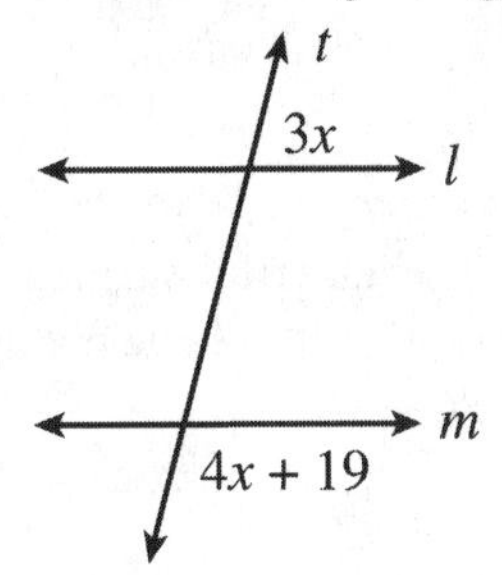

(A) 19°
(B) 23°
(C) 69°
(D) 95°
(E) 111°

ANSWER EXPLANATIONS

1. C

Recall that angles are classified, or named, based on the number of degrees in the angle. This angle has a measure of 123 degrees, which is a measure between 90 and 180 degrees. An obtuse angle has a measure in this range, so (C) is the correct answer.

Each of the other answer choices describes a different measure or range of angle measurement. Choice (A), acute angle, has a degree measure between 0 and 90 degrees. Choice (B), reflex angle, has a degree measure greater than 180. Choice (D), right angle, has a degree measure of exactly 90 degrees. Choice (E), straight angle, has a degree measure of exactly 180 degrees.

2. A

The two angles are complementary, so the sum of the measures of their angles is 90 degrees. Since you know the measure of one of the angles, subtract this amount from 90 to find the measure of the unknown angle: $90 - 65 = 25$.

The measure of the unknown angle is 25 degrees.

Choice (D) is the supplement of the angle, or the measure subtracted from 180 degrees instead of 90 degrees.

3. D

The figures in the diagram form a linear pair. In other words, the two angles are adjacent and form a straight angle together. The sum of their measures, therefore, is 180 degrees. To find the value of x, subtract the given angle of 38 degrees from the sum total of 180 degrees: $180 - 38 = 142$

The value of x is 142 degrees.

Choice (A) is the measure of the given angle. Choice (B) is the complement of the given angle. This measure added to 38 degrees is equal to 90 degrees. Choice (E) could be the result of a mathematical error in subtraction.

4. D

A strategy presented in this chapter to use when information about angle measure is given in the form of a ratio is to multiply the values in the ratio by x. The values in this ratio are 2 and 7, so multiply each by x to get the expressions $2x$ and $7x$. The two angles are complementary, so the sum of their measures is 90 degrees. To solve for the larger angle, first write an equation and solve for x.

Since the sum of the angle measures is 90 degrees, the equation can be written as: $2x + 7x = 90$

Combine like terms: $9x = 90$

Divide each side of the equation by 9 to get the x alone: $\frac{9x}{9} = \frac{90}{9}$

$x = 10$

In this problem, the smaller angle is represented by $2x$ and the larger angle by $7x$. Since this question asks for the larger angle, use the expression $7x$ and substitute $x = 10$ to find the measure of the angle: $7x = 7(10) = 70$. The measure of the angle is 70 degrees.

Choice (A) is the value of x. Choice (B) is the measure of the smaller angle. If you chose (C) or (E), you may have used 180 for the sum total of the angles. This was incorrect because the angles were complementary, not supplementary.

5. B

The first step in solving this question is to realize that the two angles are supplementary. Therefore, the sum of their angle measures is 180 degrees. The question states that one angle is 30 degrees more than twice its supplement. Using this information, let x represent the measure of the smaller angle and $2x + 30$ equal the measure of the larger angle.

Write an equation that sets the sum of these measures equal to 180 degrees:	$x + 2x + 30 = 180$
Combine like terms:	$3x + 30 = 180$
Subtract 30 from each side of the equation:	$3x + 30 - 30 = 180 - 30$
The equation becomes:	$3x = 150$
Divide each side of the equation by 3 to get x alone:	$\frac{3x}{3} = \frac{150}{3}$
	$x = 50$

Since the value of x is 50, the measure of the smaller angle is 50 degrees.

Choice (A) is used in the expression for the larger angle, but is not the measure of the larger angle. Choice (C) would be the incorrect value of x if the expression $x + 30$ was used for the larger angle, instead of the correct expression $2x + 30$. Choice (D) is twice the measure of the smaller angle, and choice (E) is the correct measure for the larger angle in the pair.

6. B

When two lines intersect, the vertical angles formed are congruent. In this figure, two lines intersect to form four angles. The vertical angles are the nonadjacent angles. Angle 2 and angle 3 are vertical, and angle 1 and the angle labeled 78 degrees are vertical. This means that the measure of angle 1 is also 78 degrees.

Choice (A) is the result of subtracting 78 degrees from 90 degrees. Choice (C) is the result of adding one to the correct measure of 78 degrees. Choice (D) is the result of subtracting 78 degrees from 180 degrees. This would be the correct measure for angle 2 or angle 3. Choice (E) is the result of subtracting 78 degrees from the sum total of all four angles which is 360 degrees.

7. E

When two lines intersect, the vertical (nonadjacent) angles formed are congruent. Expressions for angles ∠ABD and ∠CBE are given. These two angles are vertical angles, so their measures are equal. Write an equation that takes the expressions for each angle and sets them equal to each other. Then, solve the equation to find the value of x in order to find the correct answer choice.

When the two expressions are set equal to each other, the equation can be written as:	$7x - 32 = 5x + 18$
Subtract $5x$ from each side of the equation:	$7x - 5x - 32 = 5x - 5x + 18$
The equation simplifies to:	$2x - 32 = 18$
Add 32 to each side of the equation:	$2x - 32 + 32 = 18 + 32$
The equation becomes:	$2x = 50$
Divide each side of the equal sign by 2 to get x alone:	$x = 25$

This question asks for the measure of angle ∠ABD. This angle is represented by the expression $7x - 32$.

Substitute $x = 25$ into this expression to find the number of degrees in the angle:	$7x - 32$
Multiply first, then subtract:	$7(25) - 32$
	$175 - 32 = 143$

The number of degrees in angle ∠ABD is 143 degrees.

Choice (A) is the result of finding the sum of the expressions and setting it equal to 180 degrees. Choice (B) is the correct value of x, but the question asks for the number of degrees in angle ∠ABD. Choice (C) is the supplement of a 143-degree angle and would represent the two other angles in the diagram. Choice (D) could be the result of a mathematical error when solving the equation.

8. C

The equation could be set up as:	$8x - 3 = 3x + 17$
Subtract $3x$ from each side of the equation:	$8x - 3x - 3 = 3x - 3x + 17$
The equation becomes:	$5x - 3 = 17$
Add 3 to each side of the equation:	$5x - 3 + 3 = 17 + 3$
The equation is now:	$5x = 20$
Divide each side by 5 to get x alone:	$x = 4$

Since the value of x is 4, substitute this value into the expression that represents the measure of either angle. If you select the expression $8x - 3$, the measure of the angle is: $8(4) - 3 = 32 - 3 = 29$ degrees. You could check this by quickly substituting into the other expression to make sure you get the same result. $3x + 17 = 3(4) + 17 = 12 + 17 = 29$ degrees. Good work!

Choice (A) is the correct value of x, but not the measure of the angle stated in the question. Choice (B) is the value of x if the two expressions were added and set equal to 180 degrees. Choice (E) is the result if the value of x from choice (B) was substituted into either expression. Choice (D) is the complement of a 29-degree angle.

9. C

There are many patterns that emerge in a figure where two parallel lines are cut by another line. In these figures, eight angles are formed. Unless the transversal is perpendicular, four of these angles are acute in measure and all four are congruent to each other and the other four are obtuse in measure and are also congruent to each other. In addition, one of the acute angles paired with one of the obtuse angles forms a supplementary pair. The first step in this type of question is to determine if the angles given in the diagram are congruent or supplementary to each other.

In this figure, the two given angles are supplementary. One way to draw this conclusion is to use the special angle pairs studied in this chapter. One of these special angle pairs was vertical angles. In this diagram, angle 3 is vertical to the angle labeled 64 degrees. Therefore, these two angles are congruent, or have the same measure. Now, examine angle 3 and angle 1. Both of these angles are on the same side of the transversal, and both are on the interior of the parallel lines. These two angles are supplementary. The sum of their measures is 180 degrees.

To find the measure of angle 1, subtract 64 degrees from 180 degrees: $180 - 64 = 116$

The measure of angle 1 is 116 degrees.

Choice (A) is the result of subtracting 64 from 90 degrees. Choice (B) is equal to the given angle. Choice (D) is the result of a mathematical error. Choice (E) is the sum of the given angle and angle 1.

10 E

The first step in solving this problem is to determine the relationship between the two angles whose expressions are given. These two angles are on the exterior of the parallel lines, but on the same side of the transversal. Therefore, the angles are supplementary. The sum of their measures is equal to 180 degrees. Solve for x by writing an equation that sets the two expressions equal to each other.

The equation could be written as:	$3x + 4x + 19 = 180$
Combine like terms:	$7x + 19 = 180$
Subtract 15 from each side of the equation:	$7x + 19 - 19 = 180 - 19$
The equation becomes:	$7x = 161$
Divide each side by 7 to get the x alone:	$x = 23$

Since the value of x is 23, substitute into each expression to find the measure of each angle. The measure of the smaller angle is $3x = 3(23) = 69$ degrees. The measure of the larger angle is $4x + 19 = 4(23) + 19 = 92 + 19 = 111$ degrees.

Choice (A) is the value of x if the two expressions in the figure were set equal to each other and the value were made positive. Since the two angles are supplementary and not congruent, this is not the correct answer. Choice (B) is the correct value of x, but not the measure of the larger angle. Choice (C) is the measure of the smaller angle labeled $3x$ in the diagram. Choice (D) is the measure of the larger angle when $x = 19$ is substituted into the expression $4x + 19$.

CHAPTER 2

Triangles

WHAT ARE TRIANGLES?

As discussed in the first chapter, there are many basic figures in geometry that serve as the basis of other more complex concepts. In terms of geometric shapes, the triangle is one of the most basic, yet one of the most important figures. It is the polygon with the least number of sides; however, its implications on geometry and the world around us are far reaching.

Chapter 2 starts out with the basic facts about triangles: how to classify, or name, triangles, how to solve for missing angles or sides of triangles, and how to use the many properties associated with different types of triangles. Then, these properties will be expanded to solving problems using more than one triangle at a time. Even more advanced topics on the study of triangles, right triangles in particular, will continue in Chapter 12.

CONCEPTS TO HELP YOU

This chapter will start with how to classify triangles and some of the special properties associated with certain types of triangles. Even though geometry is the main focus, algebra will be used to help solve for missing sides and angles of triangles. The triangle inequality property will be explained and examples given on problems that involve the relationship between the measures of the sides of any triangle.

Classifying Triangles

A *polygon* is a simple closed curve that has line segments as sides. The *triangle* is the most basic of all polygons. Each triangle consists of three sides and three angles. The total number of degrees in the interior angles of any triangle is 180 degrees. This fact will be extremely useful when solving for missing angles and sides of triangles.

Triangles are classified based on the length of its sides and degrees in its angles. Let's start by looking at the different ways to classify triangles based on side length.

A *scalene triangle* is a triangle whose sides are each different lengths. This results in each of the angles having a different measure, also.

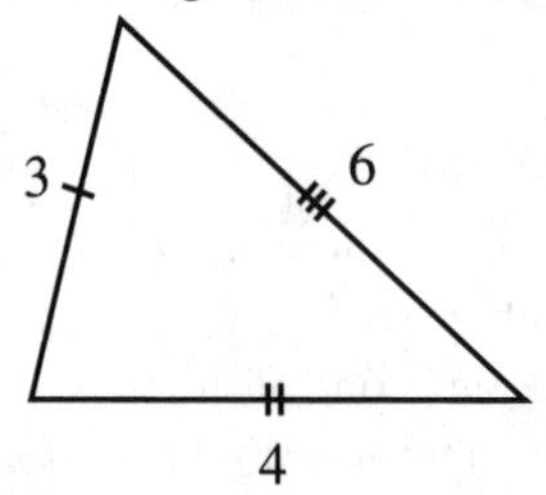

An *isosceles triangle* has two sides with the same measurement. In any isosceles triangle, the base angles across from the two congruent sides are also congruent to each other.

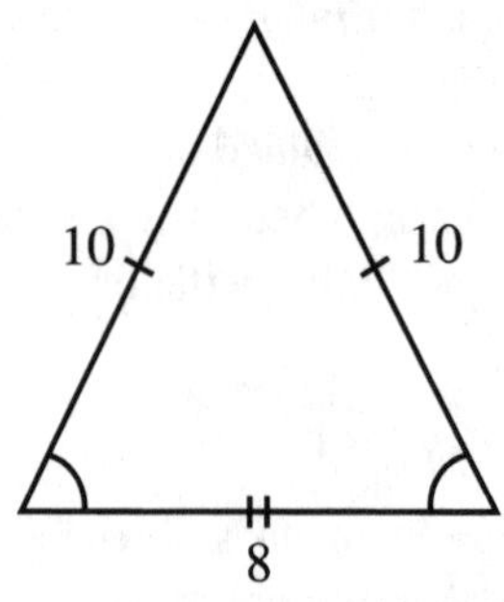

An *equilateral triangle* has three congruent sides; in other words, each side has the same measurement. Since each of the sides has the same measure, each of the angles has the same measure. Each angle is $180 \div 3 = 60$ degrees.

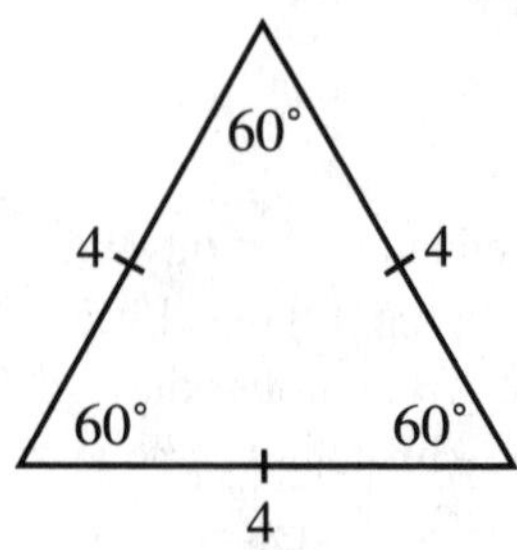

There are also three ways to classify triangles by the degrees in the angles. Refer back to Chapter 1 for further information about the types of angles described below. 

An *acute triangle* contains three angles that are acute. Recall that acute angles have a measure between 0 and 90 degrees.

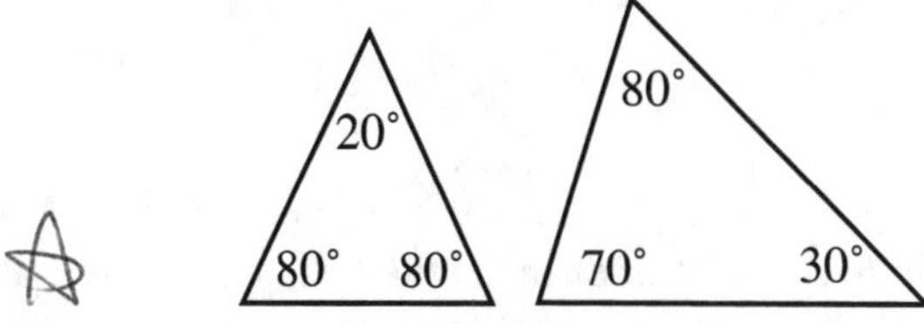

An *obtuse triangle* contains one angle that is obtuse, and two acute angles. Since an obtuse angle measures between 90 and 180 degrees, and the sum of the interior angles of a triangle is 180 degrees, there cannot be more than one obtuse angle contained in any triangle.

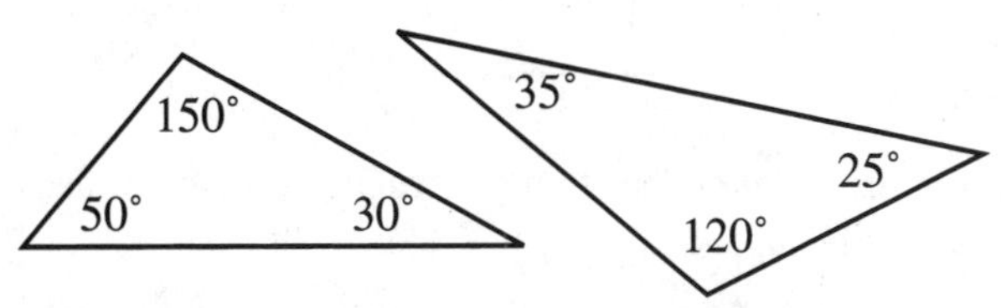

A *right triangle* is a triangle that contains exactly one right angle and two acute angles. However, the two acute angles of a right triangle are always complementary, or add to 90 degrees. Complementary angles were covered in detail in Chapter 1. The box drawn where a right angle is located indicates that the measure of the angle is 90 degrees.

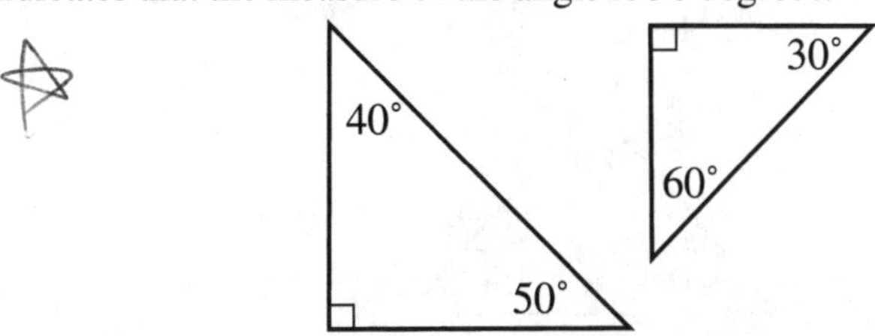

When classified by the sides and angles together, special properties emerge in some triangles. If you think of each triangle as having a first and last name, the first name refers to the sides and the last name refers to the angles. Here are a few examples.

An *isosceles right triangle* has two congruent sides, a right angle, and two acute angles that are equal in measure. Since the sum of the interior angles of a triangle is 180 degrees, and a right angle is 90 degrees, the measure of the two congruent acute angles is $(180 - 90) \div 2 = 90 \div 2 = 45$ degrees.

This is shown in the figure below.

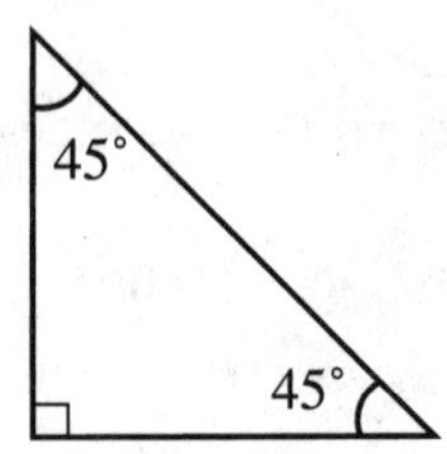

A *scalene right triangle* contains sides each with different lengths, a right angle, and two acute angles that are complementary, but not necessarily congruent.

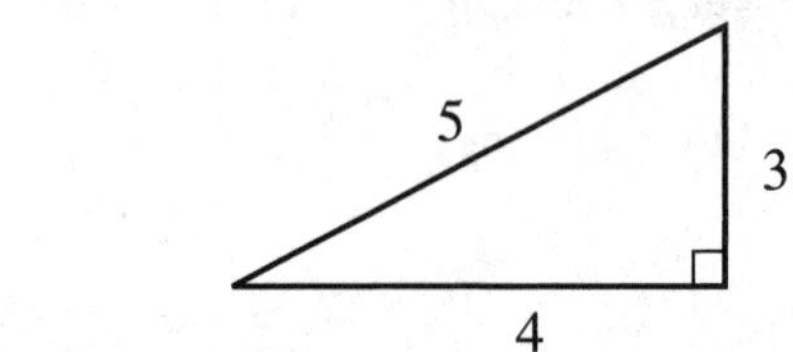

An *isosceles obtuse triangle* has two congruent sides, an obtuse angle, and two congruent acute angles.

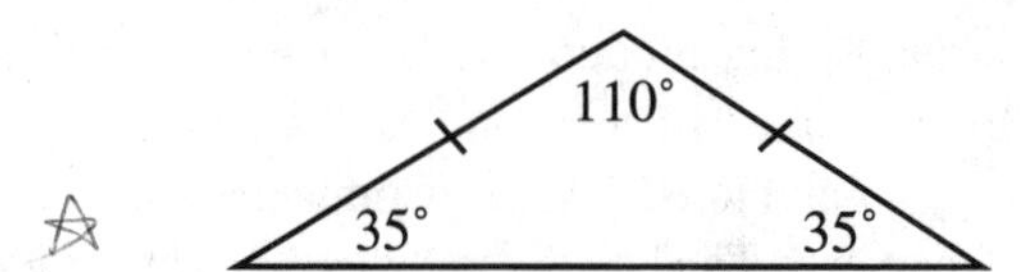

An *isosceles acute triangle* has two congruent sides and each of the angles is acute. The two acute base angles are congruent.

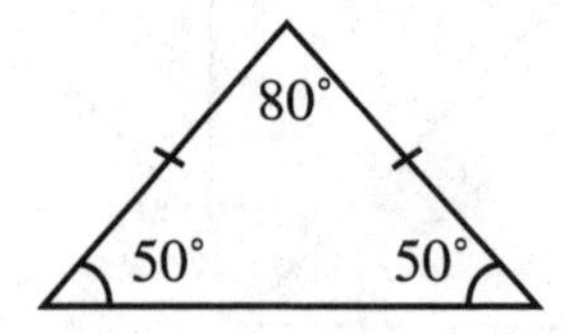

CLASSIFYING TRIANGLES

Each triangle has two parts to its name: One part describes the relationship between the sides (scalene, isosceles, equilateral) and the other describes the relationship between the angles (acute, obtuse, right).

The Triangle Inequality Property

A special property exists with regard to the length of the sides of any triangle. This property states that the sum of the two shorter sides of any triangle is greater than the length of the third side. Here are a few examples of this relationship.

1. Could 3, 5, and 7 be the lengths of the three sides of a triangle?

Yes. The two shorter sides 3 and 5 have a sum of $3 + 5 = 8$. Since 8 is greater than the longest side 7, these three measurements could form the sides of a triangle.

2. Could 2, 2, and 4 be the lengths of the three sides of a triangle?

No. The two shorter sides 2 and 2 have a sum of $2 + 2 = 4$. Since 4 is equal to the longest side, these measurements could not form the sides of a triangle.

An application of this property is finding the range of values for the third side of a triangle when two sides of a triangle are known. This range can be found by finding the sum and the difference of the two known values. For example, if two sides of a triangle are 5 and 8 then the length of the third side is between $8 - 5 = 3$ and $8 + 5 = 13$. Keep in mind that the measure of the third side is between 3 and 13 and does not include these values. Therefore, values such as 3.5, 6, 10, and 12.4 are possibilities for the length of the missing side, but the difference of 3 and the sum of 13 are not.

Another useful property related to the triangle inequality deals with the sides of the angles and the opposite sides. In any triangle, the sides opposite the angles with greater measure are longer than the sides opposite smaller angles within the same triangle. This is why the hypotenuse, or the side across from the right angle of a triangle, is always the longest side of a right triangle. More information about right triangles and their properties will be detailed in Chapter 12.

Similar Triangles

The concept of similar triangles is a very important one. Applying the properties of similar triangles is very helpful when solving some types of real-world problems.

Similar figures are figures that have the same shape, but not necessarily the same size. *Similar triangles* have corresponding sides, or the sides that line up, that are in proportion with each other. They also contain corresponding angles that are congruent. Therefore, two similar triangles have three pairs of angles with equal measure. Because of this fact, proving two triangles are similar just entails showing that two pairs of angles have the same measure. If two pairs are congruent, then the third pair will also be congruent. This is known as the *Angle-Angle Theorem*, or AA.

The key to solving similar triangle problems is to line up the corresponding parts. Lining them up correctly can be easier if the ways that the triangles are set up is understood. There are a few different ways that similar triangles are commonly presented. In one way, the triangles are side by side. This is a helpful way to view the triangles in "shadow" problems that will be discussed later in the chapter. The corresponding parts of side-by-side triangles can be lined up by thinking of one triangle as sliding over on top of the other one. Try lining up corresponding parts with these side-by-side triangles.

In the figure below, triangle ABC is similar to triangle DEF.

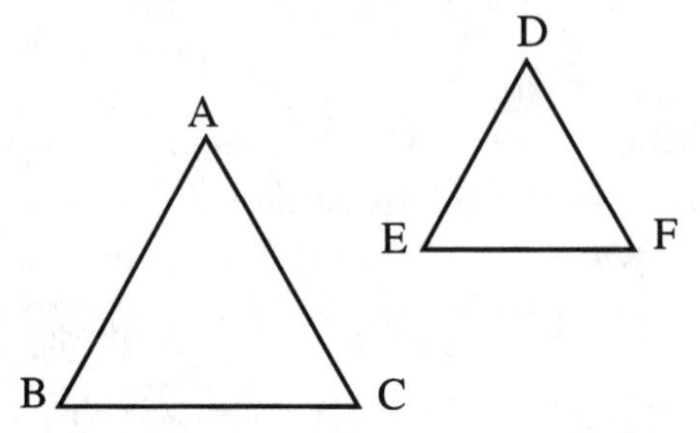

The corresponding sides are:

Side AB and side DE
Side BC and side EF
Side CA and side FD

The corresponding angles are:

Angle ∠A and angle ∠D
Angle ∠B and angle ∠E
Angle ∠C and angle ∠F

Another way that similar triangles are presented is that one triangle is located within another triangle, or the triangles overlap. This can be done in many different ways. One of these ways is shown in the examples below.

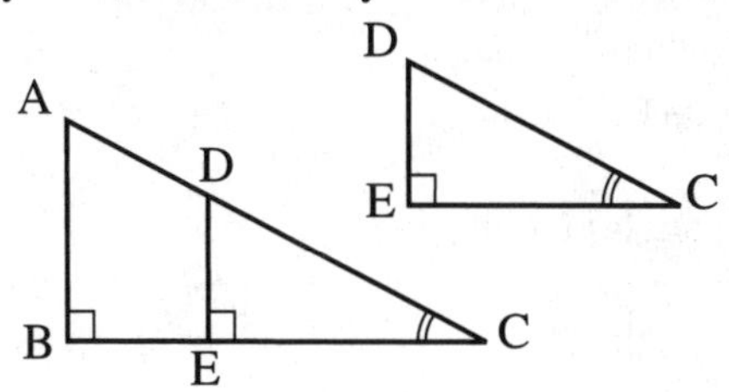

The corresponding sides are:

Side AC and side DC
Side AB and side DE
Side BC and side EC

The corresponding angles are:

Angle ∠A and angle ∠D
Angle ∠B and angle ∠E
Angle ∠C and angle ∠C (contained in both triangles)

In these examples, the corresponding parts may appear to be more difficult to identify. Be sure to line up these parts correctly before using them to solve for missing parts.

A third way could be viewed as a "bow tie," where the two triangles are facing each other. This is shown in the figure below. Use caution with this type of diagram; the triangles must actually be rotated in order to match the corresponding sides. For example, note that angle A corresponds with angle D, not angle E from the other triangle.

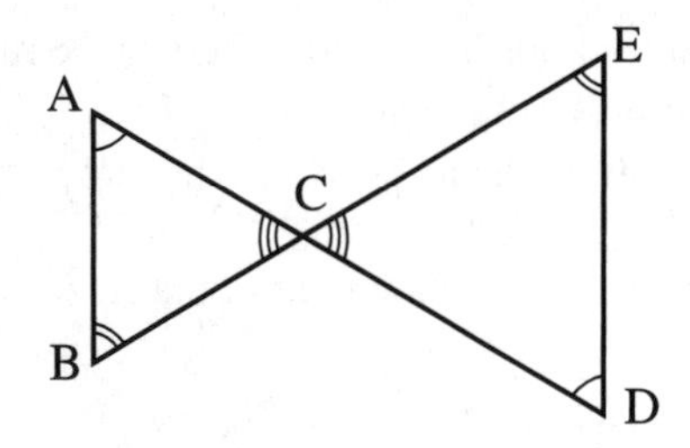

The corresponding sides are:

Side AB and side DE

Side AC and side DC

Side BC and side EC

The corresponding angles are:

Angle $\angle$A and angle $\angle$D

Angle $\angle$B and angle $\angle$E

Angle $\angle$ACB and angle $\angle$DCE

STEPS YOU NEED TO REMEMBER

SOLVING FOR THE MISSING INTERIOR ANGLE OF A TRIANGLE

Solving for missing angles within a triangle is not difficult if one fact is remembered: the sum of the interior angles of any triangle is always 180 degrees. For example, if two angles of a triangle measure 35 and 65 degrees, then the measure of the third angle is 180 minus the sum of the two known angles. The third angle would be 180 – (35 + 65) = 180 – 100 = 80 degrees.

Using Ratios to Solve for the Interior Angles of Triangles

Sometimes the information given about the angles of a triangle is given in ratio form. This concept was covered in Chapter 1 and dealt with the measures of angle pairs but is basically approached the same way with triangles. For example, suppose the three interior angles of a triangle are in the ratio 1:2:3. Find the measure of each angle.

As explained in Chapter 1, each of the numbers in the ratio can be multiplied by x to form a multiple of the ratio. Therefore, the expressions $1x$, $2x$, and $3x$ can represent the measures of the three angles.

Since the sum of the measures of the interior angles is 180,

set up the equation: $1x + 2x + 3x = 180$

Combine like terms to get the equation: $6x = 180$

Divide each side of the equation by 6 to solve for x:

$$\frac{6x}{6} = \frac{180}{6}$$
$$x = 30$$

Since the larger angles were $2x$ and $3x$, substitute $x = 30$ to find the other angles.

$$2x = 2(30) = 60$$
$$3x = 3(30) = 90$$

The three angles of the triangle are 30, 60, and 90 degrees.

Solving for the Exterior Angle of a Triangle

With many math problems, there are a number of different ways to solve them. This concept is true when trying to find the exterior angle of a triangle. One way is to use the given measurements of the interior angles and make a linear pair with the exterior angle. A linear pair, covered in Chapter 1, is a pair of adjacent angles whose sum is 180 degrees. If the measure of the interior angle of the linear pair is found, then the measure of the exterior angle is 180 – (measure of the interior angle).

To find the measure of the exterior angle x below, first use the two known interior angles to find the third interior angle. Since the two angles are 75 and 40, this is a sum of 115. Subtract this sum from 180 to find the measure of the third interior angle: 180 – 115 = 65 degrees. This angle forms a linear pair with the exterior angle. Subtract 180 – 65 = 115, which is the measure of the exterior angle.

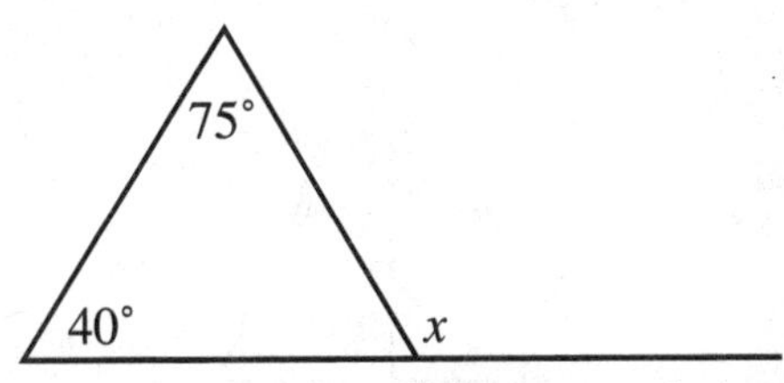

Another method to use when seeking the measure of an exterior angle is a special relationship: the measure of an exterior angle is equal to the sum of the two *remote interior angles.* So, what are the remote interior angles? These are the two interior angles that are not adjacent to the exterior angle, or the two angles that are farthest away from the exterior angle. In the figure above, the two remote interior angles were the angles labeled 75 and 40 degrees. Their sum is 115, which is also the measure of the exterior angle. When remembered, this relationship can be a big time-saver when solving for exterior angles.

Properties of Isosceles Triangles

Isosceles triangles were introduced earlier in this chapter, and have numerous properties that can be very helpful when solving for missing parts. Each of the angles of isosceles triangles has a special name. The two congruent angles across from the congruent sides are known as the *base angles*, and the third angle is known as the *vertex angle.* Therefore, even if only one angle of an isosceles triangle is known, the other angles may be solved for. If the vertex angle is known, subtract its measure from 180 and divide the result by 2, since there are two base angles. For example, if the vertex angle of an isosceles triangle measures 50 degrees, then the measure of each base angle is $(180 - 50) \div 2 = 130 \div 2 = 65$ degrees.

If one of the base angles is known, double its value (since there are two base angles) and subtract the result from 180 to find the vertex angle. For example, if a base angle measures 40 degrees, double this value to 80 degrees to represent both of the base angles. Then, subtract this amount from 180 to find the measure of the vertex angle. The measure of the vertex angle would be $180 - 80 = 100$ degrees.

Another pattern appears when the altitude in an isosceles triangle is drawn from the vertex angle to the opposite side of the triangle. This line creates two congruent right triangles and is shown in the figure below.

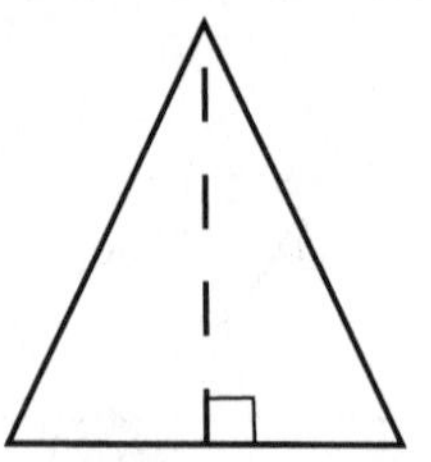

When the altitude is drawn in an equilateral triangle, it forms two congruent right triangles. The measures of each angle in this right triangle are 30, 60, and 90 degrees. This special right triangle will be discussed in detail in Chapter 12.

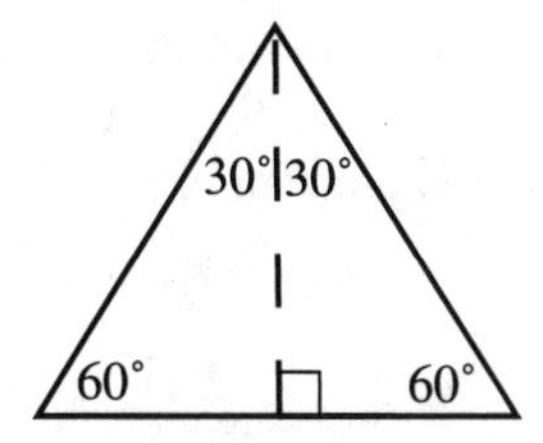

Setting Up Similar Triangle Problems

An important concept to understand when solving similar triangle questions is lining up corresponding parts. These are the parts of the two triangles that would match up if the two triangles were placed one on top of the other. Go back to the section in this chapter labeled Similar Triangles under Concepts to Help You for additional review on lining up corresponding parts.

Once these corresponding parts are located, these measurements can be used in proportions to solve for missing sides. A *proportion* consists of two ratios set equal to each other. Using two sides of each triangle involved in the problem creates these ratios. To set up these ratios with corresponding parts, use labels to help you. For example, consider the two similar triangles below.

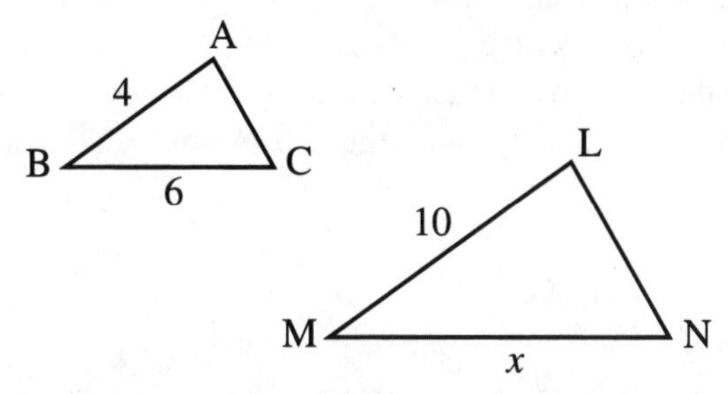

The corresponding sides are as follows:

Side AB corresponds with side LM

Side BC corresponds with side MN

Side CA corresponds with side NL

Using this information, set up a proportion using a pair of corresponding sides where one side is unknown, and another pair of sides where both measures are known. Use labels, as shown below, to ensure that the corresponding parts are in the correct place in the proportion.

$$\frac{\text{Side of small triangle}}{\text{Base of small triangle}} = \frac{\text{Side of small triangle}}{\text{Base of small triangle}}$$

Now, substitute the values for each side, using x for the missing side. $\frac{4}{6} = \frac{10}{x}$

Cross multiply the values in the proportion. $4 \cdot x = 6 \cdot 10$

$4x = 60$

Divide by 4 to get x alone. $\frac{4x}{4} = \frac{60}{4}$

The measure of side MN is 15 units. $x = 15$

There are also many real-world applications of similar triangles. A frequently used problem type is called a "shadow" problem. This type of problem uses similar triangles and indirect measurement to find the measure of objects that would be difficult to measure using conventional means. Consider the example below:

> A tree in the park has a shadow that measures 12 feet. At the same time, a woman five feet tall standing next to the tree has a shadow that is three feet long. What is the height of the tree?

The height of the tree could be difficult to find if you tried to measure it using a ladder and tape measure. However, similar triangles can be used in this case and the problem becomes a lot easier. The first step is to make a diagram showing the woman and the tree with their shadows. Label each known measurement and label the measure you are looking for, which is the height of the tree, with an x. The figure below is a possible diagram for this scenario.

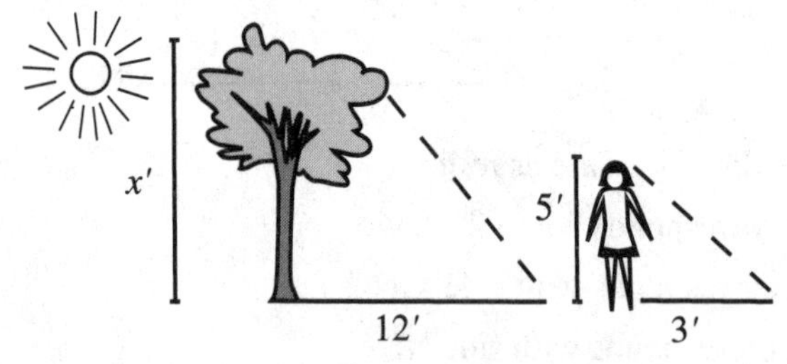

Note that a dotted line is drawn from the top of the tree to the end of its shadow, and from the top of the woman's head to the end of her shadow, to create the similar triangles.

Now, line up corresponding parts from the diagram in a proportion. One possible proportion is:

$$\frac{\text{Height of woman}}{\text{Length of woman's shadow}} = \frac{\text{Height of tree}}{\text{Length of tree's shadow}}$$

Substitute the values from the problem into the proportion. Remember to use x to represent the height of the tree in the proportion: $\frac{5}{3}=\frac{x}{2}$

Cross multiply the values in the proportion: $3 \cdot x = 5 \cdot 12$

$3x = 60$

Divide each side by 3 to get x alone: $\frac{3x}{3}=\frac{60}{3}$

$x = 20$

Therefore, the height of the tree is 20 feet. This was a much more efficient, and safer, way of finding the height of the tree than using a ladder!

STEP-BY-STEP ILLUSTRATION OF THE FIVE MOST COMMON QUESTION TYPES

This section of the chapter identifies some of the most common question types concerning triangles and provides answers with detailed explanations.

Question 1: Classifying a Triangle

Which of the following best describes the triangle below?

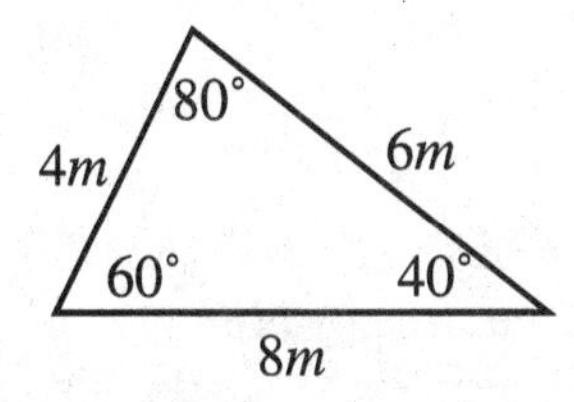

(A) Scalene right

(B) Isosceles right

(C) Isosceles acute

(D) Scalene acute

(E) Equilateral acute

Triangles are classified based on their angle and side measurements. Each triangle may have two words associated with it. One of the terms describes the sides and would be equilateral if all of the sides are congruent, isosceles if two sides are congruent, and scalene if each side has a different measure. The other term describes the angle measure. The term acute means all three angles are acute, obtuse means that one angle is obtuse and the other two angles are acute, and right means that one angle is right and the other two acute angles are complementary.

The triangle in the figure above has three different measures for its sides. This makes the triangle scalene. The angle measures are all different and each is less than 90 degrees. This makes the triangle acute. **Therefore, the correct answer choice is (D), scalene acute.** Choice (A) would have all sides with a different length and contain a right angle. Choice (B) would have two congruent sides and a right angle. Choice (C) would have two congruent sides and contain three angles less than 90 degrees. Choice (E) would have three congruent sides and three 60-degree angles.

Question 2: Finding the Missing Interior Angle of a Triangle with Ratios

The three angles of a triangle are in the ratio 2:3:4. What is the measure of the largest angle of this triangle?

(A) 20º

(B) 40º

(C) 60º

(D) 80º

(E) 100º

The three angles of a triangle always have a sum of 180 degrees. For this problem, the ratio of the three angles is given. One strategy presented in this section is to make each value in the ratio a multiple of x. To do this, multiply each number in the ratio by x. Then, the smallest angle would be represented by $2x$, the next largest angle by $3x$, and the largest angle by $4x$.

Next, use algebra by setting up an equation that shows the relationship of the three angles.

Since the sum is 180 degrees, set up the equation: $2x + 3x + 4x = 180$

Combine like terms. $9x = 180$

Divide each side of the equation by 9 to get x alone: $\frac{9x}{9} = \frac{180}{9}$

$x = 20$

The expression $4x$ represents the largest angle of the triangle.

Substitute $x = 20$ to find the measure of this angle: $4x = 4(20) = 80$

The largest angle is 80°, **which is choice (D).** This question asks for the largest angle of the triangle, not the measure of x, so be sure not to choose choice (A). Choice (B) is the measure of the smallest angle of the triangle, and choice (C) is the measure of the next largest angle.

Question 3: Finding the Missing Exterior Angle of a Triangle

What is the measure, in degrees, of the exterior angle ACD in the figure below?

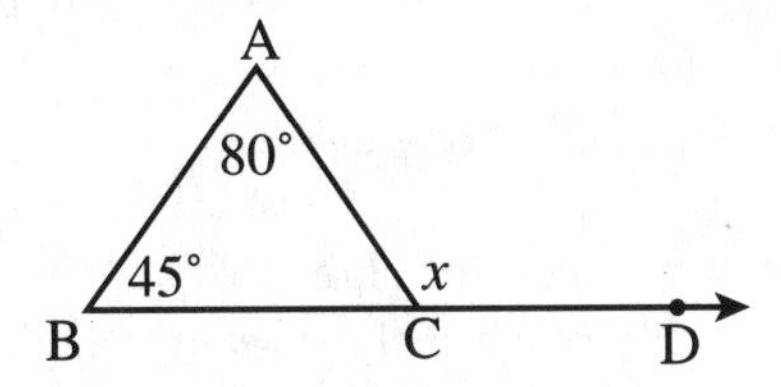

(A) 45°

(B) 55°

(C) 80°

(D) 115°

(E) 125°

As discussed earlier in this chapter, there is more than one way to find the exterior angle of a triangle. The first way is to use the adjacent interior angle. An exterior angle of a triangle forms a linear pair with the adjacent interior angle. In other words, the exterior angle and the interior angle next to it form a 180 degree, or straight, angle. If the measure of the interior angle is known, then this value can be subtracted from 180 to find the measure of the exterior angle. In this figure, the angle adjacent to the exterior angle is not known, but the measure of the other two interior angles is. These angles measure 80 and 45 degrees. Since the sum of the interior angles of a triangle is 180 degrees, subtract the sum of the two known angles from 180 to find the third angle.

In this case, the measure of the third angle is: $180 - (80 + 45)$
$180 - 125 = 55$

Using this information, the measure of the exterior angle is then equal to $180 - 55 = 125°$ **which is choice (E).**

Another method for finding the measure of an exterior angle is to use the following property:

> *The measure of an exterior angle of a triangle is equal to the sum of the remote interior angles.*

Basically, this property says that if you add the measures of the two angles in the triangle that are not adjacent, or next to, the exterior angle, their sum is equal to this exterior angle. Try this property with the question above. The two known interior angles are 80 and 45 degrees, and neither is adjacent to the exterior angle in the figure. The sum of these two angles is $80 + 45 = 125$ degrees. This is also the measure of the exterior angle ACD.

If you chose choices (A) or (C), these are the measures of the two known interior angles in the figure. Choice (B) is the measure of the interior angle adjacent to exterior angle ACD.

Question 4: Using the Triangle Inequality Property

Two sides of a triangle measure 12 m and 3 m. Which of the following could represent the measure of the third side?

(A) 3 m

(B) 6 m

(C) 9 m

(D) 13 m

(E) 15 m

The triangle inequality property is very useful to know when working with the sides of a triangle. It can help you to easily eliminate impossible answers, when used correctly. As stated earlier in this chapter, this property says that the sum of the two smaller sides of a triangle must be greater than the third side. Another way to look at this property is that if the lengths of two sides of a triangle are known, then the third side must be a measure between the sum and the difference of those sides. In this case, the two known sides are 12 and 3. The difference of these values is $12 - 3 = 9$. The sum of the sides is $12 + 3 = 15$. Therefore, the third side must be between (and not include) 9 and 15.

Answer choices (A), (B), and (C) are too small. Even with choice (C), the two smaller sides would be 3 and 9. Since $3 + 9 = 12$, this is equal to the third side, not greater than it. Lengths of 3, 9, and 12 would not form a triangle. Choice (E) is too large. **Only choice (D), 13, could be the length of the third side.** To check this, the three sides would be 3, 12, and 13. The two smaller sides, 3 and 12, have a sum of 15, which is greater than 13. The lengths 3, 12, and 13 could form the sides of a triangle.

Question 5: Solving Problems with Similar Triangles

A 12-foot tall flagpole is located next to a school building. If the shadow cast by the flagpole is 16 feet at the same time the shadow of the building is 24 feet, what is the height of the building?

(A) 16 feet

(B) 18 feet

(C) 20 feet

(D) 24 feet

(E) 32 feet

The key to solving any problems involving similar triangles is to recognize and line up the corresponding parts between the triangles that can be drawn. Recall that corresponding parts are the parts that would line up if the triangles were placed one on top of the other.

A diagram of the situation in the question could be represented as the figure below.

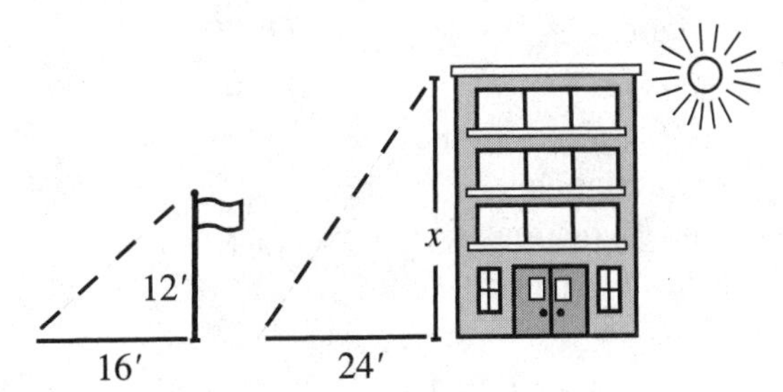

When a dotted line is drawn to connect the top of each object with the end of its shadow, similar triangles are formed.

The first step is to line up the corresponding parts from the diagram in a proportion. One possible proportion is:

$$\frac{\text{Height of flagpole}}{\text{Length of flagpole's shadow}} = \frac{\text{Height of school}}{\text{Length of school's shadow}}$$

Substitute the values from the problem into the proportion. Remember to use x to represent the height of the school in the proportion:

$$\frac{12}{16} = \frac{x}{24}$$

Cross multiply the values in the proportion:

$$16 \cdot x = 12 \cdot 24$$
$$16x = 288$$

Divide each side by 16 to get x alone:

$$\frac{16x}{16} = \frac{288}{16}$$
$$x = 18$$

The correct answer is choice (B), 18 feet. Choices (A) and (D) are the lengths of the shadows of the objects in the problem but are not the correct height of the school. You may have selected choice (C) if you thought that the length of the shadow was four more than the height of each object. Choice (E) is the result of setting up an incorrect proportion using the values in the problem.

CHAPTER QUIZ

1. A triangle has angles that measure 95°, 45°, and 40°. Which of the following best describes this triangle?

   (A) Scalene right

   (B) Scalene obtuse

   (C) Isosceles obtuse

   (D) Isosceles right

   (E) Scalene acute

2. If two angles of a triangle measure 64 and 32 degrees, what is the measure of the third angle?

   (A) 32°

   (B) 64°

   (C) 84°

   (D) 94°

   (E) 96°

3. The three angles of a triangle are in the ratio 4:5:9. What is the measure of the smallest angle?

   (A) 10°

   (B) 18°

   (C) 40°

   (D) 50°

   (E) 90°

4. What is the measure, in degrees, of the exterior angle ∠LMP in the figure below?

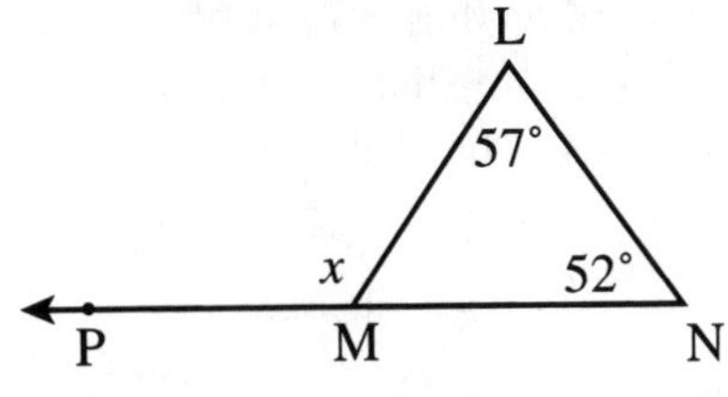

(A) 52°
(B) 57°
(C) 71°
(D) 109°
(E) 180°

5. If the measure of the vertex angle of an isosceles triangle is twice the measure of a base angle, what is the measure of the vertex angle of this triangle?
(A) 35°
(B) 45°
(C) 70°
(D) 90°
(E) 135°

6. If two sides of a triangle measure 4 m and 7 m, which of the following could be the measure of the third side?
(A) 2 m
(B) 3 m
(C) 10 m
(D) 11 m
(E) 14 m

7. Which answer choice could not represent the length of three sides of a triangle?
(A) 2, 2, 3
(B) 2, 2, 5
(C) 6, 8, 10
(D) 4, 4, 4
(E) 7, 10, 12.5

8. Which segment is the longest side of the triangle below?

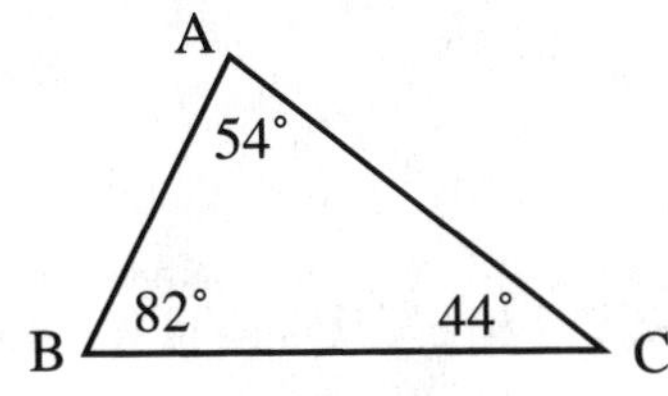

(A) Side AB
(B) Side BC
(C) Side AC
(D) Sides AB and AC
(E) All sides are congruent.

9. What is the measure of side AB in the triangle below?

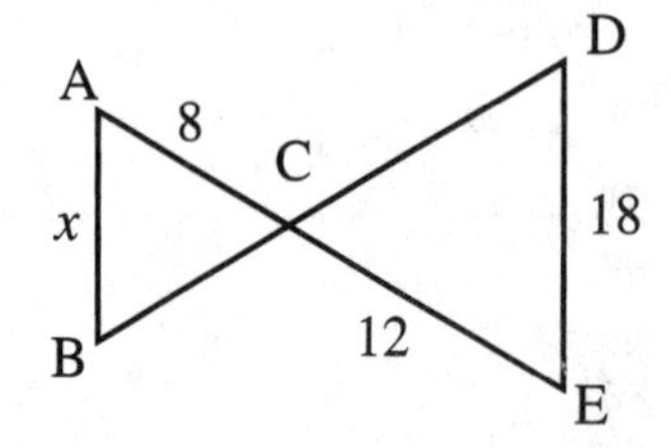

(A) 1.5

(B) 5.3

(C) 10

(D) 12

(E) 14

10. A 6-foot tall football player is standing on a field next to the goal post. If the player casts a shadow that is 5 feet long at the same time that the goal post casts a shadow that is 25 feet long, how tall is the goal post?

(A) 11 feet

(B) 20.83 feet

(C) 26 feet

(D) 30 feet

(E) 31 feet

ANSWER EXPLANATIONS

1. B

The triangle in this question has three different measures for its angles. This makes each of the sides of the triangle different measures and means that the triangle is scalene. One of the angles is greater than 90 degrees, so the triangle is obtuse. Therefore, the correct answer choice is (B), scalene obtuse.

Choice (A) would have all sides with a different length and contain a right angle. Choice (C) would have two congruent sides and one obtuse angle. Choice (D) would have two congruent sides and contain one angle equal to 90 degrees. Choice (E) would have three sides with different measures and all angles less than 90 degrees.

2. C

The sum of the interior angles of any triangle is 180 degrees. In this question, two of the interior angles of a triangle are given. Since you are looking for the measure of the third angle in the triangle, find the sum of the two known angles and subtract this sum from 180 degrees.

The two known angles are 64 and 32. The sum of the measures of these angles is $64 + 32 = 96$ degrees. Subtract this amount from 180 degrees to find the measure of the third angle: $180 - 96 = 84$. The measure of the third angle is 84 degrees.

Choice (A) is the difference between the two given angles, as well as the measure of the smaller angle. Choice (B) is the measure of the larger given angle. Choice (D) is the result of a mathematical error. Choice (E) is the sum of the two given angles.

3. C

The three angles of a triangle always have a sum of 180 degrees. Multiply each number in the ratio by x. Then, the smallest angle would be represented by $4x$, the next largest angle by $5x$, and the largest angle by $9x$.

Since the sum of the three angles is 180 degrees, set up the equation: $4x + 5x + 9x = 180$

Combine like terms: $18x = 180$

Divide each side of the equation by 18 to get x alone: $\frac{18x}{18} = \frac{180}{18}$

$x = 10$

The expression $4x$ represents the smallest angle of the triangle.

Substitute $x = 10$ to find the measure of this angle: $4x = 4(10) = 40$

The measure of the smallest angle is 40 degrees.

This question asks for the smallest angle of the triangle, not the measure of x, so be sure not to choose choice (A). Answer choice (B) is the sum of the values of the ratio. Choice (D) is the measure of the next largest angle in the triangle. The largest angle is 90 degrees, which is choice (E).

4. D

Since the sum of the interior angles of a triangle is 180 degrees, subtract the sum of the two known angles from 180 to find the third angle. In this case, the measure of the third angle is:

$180 - (57 + 52)$
$180 - 109 = 71$ degrees

Using this information, the measure of the exterior angle is then equal to $180 - 71 = 109$ degrees, which is choice (D).

Another method for finding the measure of an exterior angle is to apply the rule that the measure of an exterior angle of a triangle is equal to the sum of the remote interior angles.

The two known interior angles are 57 and 52 degrees, and neither is adjacent to the exterior angle in the figure. The sum of these two angles is $57 + 52 = 109$ degrees. This is also the measure of the exterior angle $\angle LMP$.

If you chose choice (A) or (B), these are the measures of the two known interior angles in the figure. Choice (C) is the measure of the interior angle adjacent to exterior angle $\angle LMP$. Choice (E) is the sum of the three interior angles of the triangle, or the sum of the exterior angle and the adjacent interior angle.

5. D

In any isosceles triangle, there are two base angles and one vertex angle. The base angles are across from the two congruent sides, so the base angles are also congruent. The vertex angle is across from the third side and can have a measure different from the two base angles. Since the measure of the vertex angle is twice the measure of a base angle, let x = the measure of a base angle, and let $2x$ = the measure of the vertex angle. The three angles of a triangle have a sum of 180 degrees. Use this fact to write an equation to find the measure of the angles.

Add each expression for the three angles of the triangle:	$x + x + 2x = 180$
Combine like terms.	$4x = 180$
Divide each side of the equation by 4.	$\frac{4x}{4} = \frac{180}{4}$
The value of x is:	$x = 45$

Since x is equal to 45, the measure of each base angle is 45 degrees. This question is looking for the measure of the vertex angle, so $2x = 2(45) = 90$ degrees.

Choice (B) is the measure of one of the base angles. Choice (E) is the sum of one base angle and the vertex angle.

6. C

In any triangle, the sum of the lengths of the two smaller sides must be greater than the length of the third side. In order to find possible measurements for the third side of the triangle, find the sum and difference of the two known sides. The two known sides are 4 and 7, so the difference is $7 - 4 = 3$ and the sum is $7 + 4 = 11$. The third side must be between 3 m and 11 m. The only answer choice in this range is choice (C), 10 m.

Choices (A) and (B) are too small to be in this range. The measures in choices (D) and (E) are too large to be the measure of the third side.

7. B

In any triangle, the sum of the lengths of the two smaller sides must be greater than the length of the third side. This relationship is true for each answer choice, except for choice (B). In this choice, the values for the sides of the triangle are 2, 2, and 5. If the two smaller sides are added together, the sum is $2 + 2 = 4$. This sum is less than the measure of the third side, since $4 < 5$. Thus, the measures in choice (B) could not represent the sides of a triangle.

Choice (A) is true because $2 + 2 = 4$, which is greater than 3. Choice (C) is true because $6 + 8 = 14$, which is greater than 10. Choice (D) is true because $4 + 4 = 8$, which is greater than 4. Choice (E) is true because $7 + 10 = 17$, which is greater than 12.5.

8. C

An important relationship involving triangles is that the longest side of a triangle is across from the largest angle in the triangle. In the figure, the largest angle is angle $\angle B$. This angle measures 82 degrees. Since this is the largest angle, the longest side is across from this angle. This is side AC.

Choice (A) is the smallest side of the triangle because it is across from the smallest angle. Choice (B) is longer than side AB, but shorter than side AC since the angle across from it has a measure that is between the other two angles. Choice (D) is not correct because two of the angles are not congruent, and the triangle is not an isosceles triangle. All sides are not congruent because all angles are not the same measure, so choice (E) is incorrect.

9. D

The first step is to line up the corresponding parts from the diagram in a proportion.

One possible proportion is: $\dfrac{\text{Side AB}}{\text{Side AC}} = \dfrac{\text{Side DE}}{\text{Side EC}}$

Substitute the values from the problem into the proportion.

Remember to use x to represent the missing side in the proportion: $\dfrac{x}{8} = \dfrac{18}{12}$

Cross multiply the values in the proportion: $12 \bullet x = 8 \bullet 18$

$12x = 144$

Divide each side by 12 to get x alone: $\dfrac{12x}{12} = \dfrac{144}{12}$

$x = 12$

Choice (A) is the result of dividing one side of the larger triangle by the corresponding side of the smaller triangle. Choice (B) is the result of setting up an incorrect proportion using the given information. Choice (C) is the result of adding two to the known side of the smaller triangle.

10. D

The best way to solve this question is to start by drawing a picture of the situation. An example diagram is drawn below. Connect the top of the player's head with the end of his shadow and the top of the goal post with the end of its shadow. Now, the similar triangles can be seen.

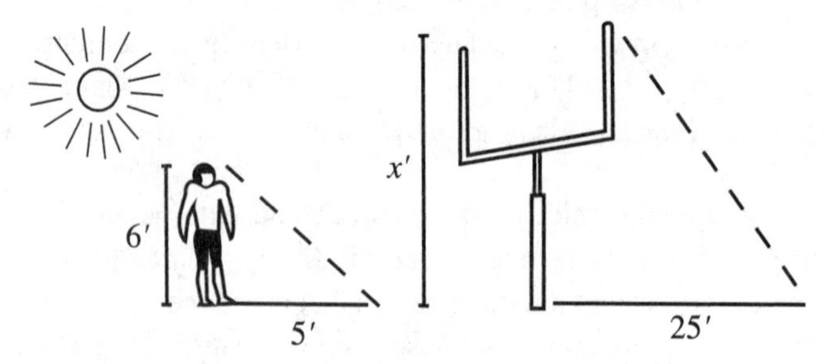

The next step is to line up the corresponding parts from the diagram in a proportion.

One possible proportion is:

$$\frac{\text{Height of player}}{\text{Length of player's shadow}} = \frac{\text{Height of goalpost}}{\text{Length of goalpost's shadow}}$$

Substitute the values from the problem into the proportion. Remember to use x to represent the height of the goal post in the proportion: $\frac{6}{5} = \frac{x}{25}$

Cross multiply the values in the proportion: $5 \cdot x = 6 \cdot 25$

$5x = 150$

Divide each side by 5 to get x alone. $\frac{5x}{5} = \frac{150}{5}$

$x = 30$

The correct answer is choice (D), 30 feet. Choice (A) is the sum of the height of the player and the shadow of the player. Choice (B) is the result of setting up an incorrect proportion using the values in the problem. You may have selected choice (C) if you thought you should add 20 to the height of the player. Choice (E) is the result of adding the height of the player with the length of the shadow of the goal post.

CHAPTER 3

Quadrilaterals and Other Polygons

WHAT IS A POLYGON?

In Chapter 2, you were introduced to the triangle. A triangle is a three-sided polygon. This chapter will expand on the other basic two-dimensional figures called *polygons*. A polygon is a closed figure made up of sides that are line segments. A *closed figure* is one in which the line segments join at the endpoints, with no opening to the interior.

This chapter will walk you through classification of a polygon, the special classifications of quadrilaterals, and working with the sides and angles of polygons.

CONCEPTS TO HELP YOU

This section will define the core knowledge needed for dealing with polygons. First, you will learn how polygons are classified according to the number of sides. Then we will move onto the common polygon called the quadrilateral. Just as with triangles, there is a classification system set up just for quadrilaterals. We'll finish up with some statements of facts concerning the interior and exterior angles of a polygon, as well as facts about the sides and diagonals of quadrilaterals.

Classifying Polygons

Polygons are classified according to the number of sides that are used to create it. Use the chart below to classify the more common polygons.

Polygon Classification

NUMBER OF SIDES	NAME
3	Triangle
4	Quadrilateral
5	Pentagon
6	Hexagon
7	Heptagon
8	Octagon
9	Nonagon
10	Decagon

Another useful definition is the fact that a *regular polygon* has all sides and all angles that are congruent, or equal in measure. A *diagonal* of a polygon is a line segment in the interior that connects any two nonadjacent vertices. A triangle has no diagonals; each vertex is adjacent to the other two vertices. The following figure illustrates the diagonals of some of the common polygons: the quadrilateral, pentagon, and hexagon.

The Diagonals of Three Polygons

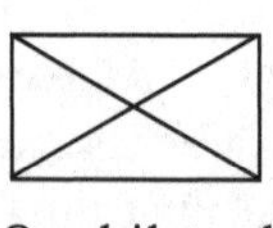

Quadrilateral Pentagon Hexagon

The triangle is probably the most common and basic polygon and is covered in Chapter 2. Let's now move onto the next most common polygon—the quadrilateral.

Classifying Quadrilaterals

A *quadrilateral* is a four-sided polygon. You are no doubt very familiar with the terms *rectangle* and *square*. These are actually specific types of quadrilaterals that fall under a broader classification. Some quadrilaterals have either one or two pairs of parallel sides. If there is exactly one set of parallel sides, it is called a *trapezoid*. Some examples of trapezoids are shown below.

Examples of Trapezoids

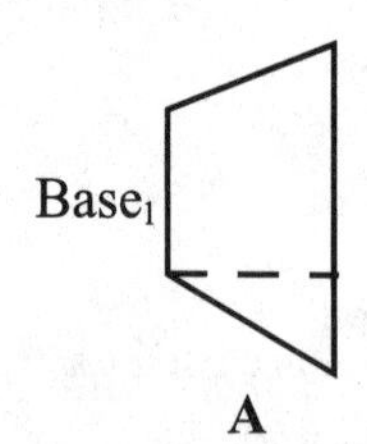

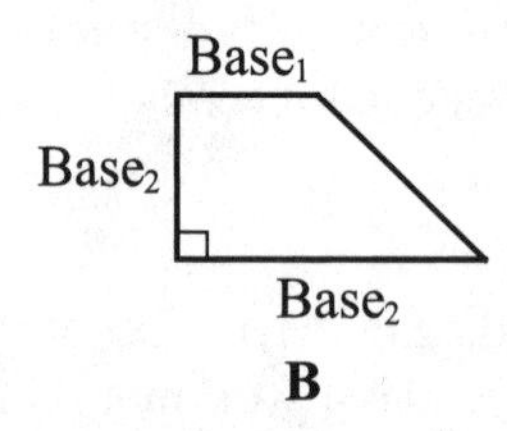

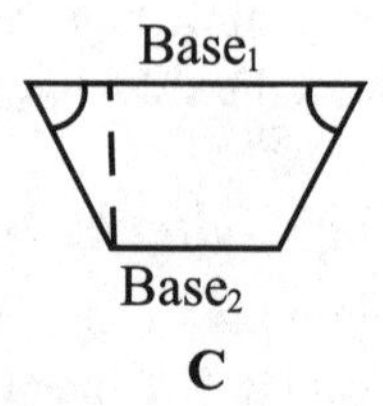

The parallel sides are called *bases*, traditionally called b_1 and b_2. The *height* is a segment perpendicular to the bases. In trapezoid B above, the height is also the left-hand side. If the two nonparallel sides are equal in measure, then the trapezoid is an *isosceles trapezoid*, shown as trapezoid C above. For an isosceles trapezoid, the base angles and the diagonals are congruent.

When there are two sets of parallel sides, the quadrilateral is classified as a *parallelogram*.

In a parallelogram, the opposite sides and opposite angles are congruent. In any parallelogram, the diagonals divide each other in half, or bisect each other.

Under the category of parallelogram are two more specific types. A *rhombus* has all sides congruent. Two examples of rhombuses are parallelograms (B) and (D) above. The diagonals of a rhombus not only bisect each other, but they are also perpendicular. A *rectangle* is a parallelogram with four right angles, shown as figures (C) and (D) above. The diagonals of a rectangle bisect each other and are also of equal measure (congruent). Notice that the *square*, figure (D) above, falls under both categories of rhombus and rectangle.

The following diagram brings these classifications together to help your understanding.

Quadrilateral Classification

Quadrilateral

No parallel sides

Trapezoid
One set of parallel sides

Isosceles Trapezoid
Nonparallel sides congruent

Parallelogram
Two sets of parallel sides

Rectangle
Four 90° angles

Rhombus
All sides congruent

Square

Properties of Quadrilaterals

The following chart summarizes the special properties of quadrilaterals. It is helpful to know these facts in order to solve a variety of problems.

Properties of Quadrilaterals

	Quadrilaterals	Trapezoid	Isosceles Trapezoid	Parallelogram	Rectangle	Rhombus	Square
Has four sides	*	*	*	*	*	*	*
Has exactly one pair of parallel sides		*	*				
Has two pairs of parallel sides				*	*	*	*
Has four 90° angles					*		*
Opposite sides are congruent				*	*	*	*
All sides are congruent						*	*
Diagonals bisect each other				*	*	*	*
Diagonals are congruent			*		*		*
Diagonals are perpendicular						*	*
Base Angles are congruent			*				

The Angles of a Polygon

You learned in Chapter 2 that the sum of the measure of the interior angles in a triangle is always 180°. Now, study the following figure which illustrates polygons with triangles showing angle measure.

Polygons Broken into Triangular Areas

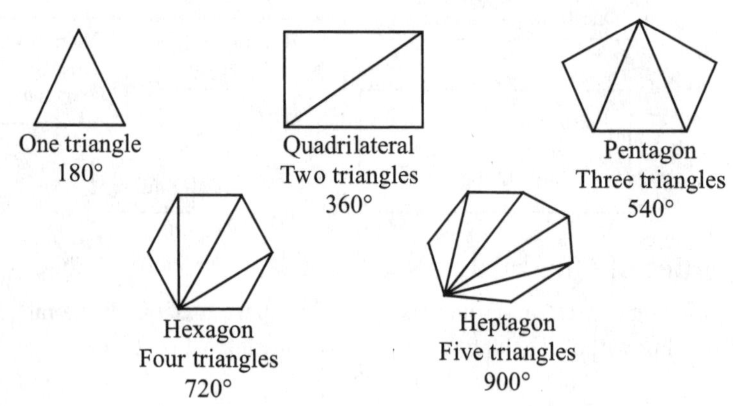

Number of Sides	Sum of the Degree Measure	Total Degree Measure
3	$180° \cdot 1$	180°
4	$180° \cdot 2$	360°
5	$180° \cdot 3$	540°
6	$180° \cdot 4$	720°
7	$180° \cdot 5$	900°

Notice that a quadrilateral can be made up of two triangles; therefore the sum of the interior angles is 180° + 180° = 360°. A pentagon can be made up of three triangles; therefore the sum of the interior angles is 3 • 180° = 540°. If you follow the pattern, you are led to an important fact about the sum of the interior angles in a polygon, which is:

If n represents the number of sides in a polygon, then the sum of the measures of the interior angles of the polygon is $180(n - 2)$. Try the formula to verify that it works.

STEPS YOU NEED TO REMEMBER

The following steps will assist you in tackling geometry problems dealing with polygons.

How to Determine the Interior and Exterior Angles of a Polygon

You saw in the last section how to determine the sum of all of the interior angles in a polygon. Use this fact when answering questions that ask you to find the measure of a missing angle. For example, given the pentagon below, find the measure of n:

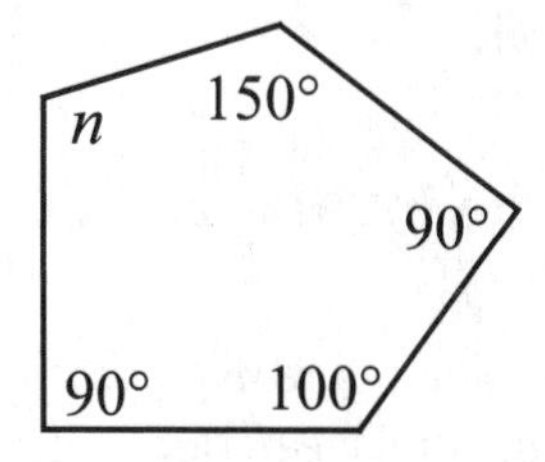

Knowing that the sum of the angles is 180(5 – 2) = 180(3) = 540, add up all given measures and then subtract it from 540: 540 – (150 + 90 + 100 + 90) = 540 – 430 = 110°.

Another application is to find the measure of an angle in a regular hexagon. Because you are told the polygon is regular, each angle is congruent. So just divide the sum by the number of sides. The sum of the interior angles in a hexagon is $180(6 - 2) = 180(4) = 720°$. Each angle will be $720 \div 6 = 120°$.

Recall from Chapter 1 that a linear pair is two adjacent angles that sum to 180° and form a straight line. In Chapter 2, you found that an *exterior angle* to a triangle forms a linear pair with one of the interior angles. This is true of exterior angles in any polygon. To find the measure of an exterior angle, subtract the measure of the interior angle adjacent to it from 180°. So if a question asks you for the value of x below, $x = 140°$, because $180 - 40 = 140°$.

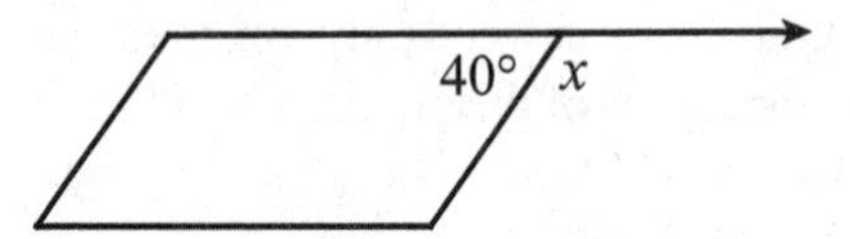

Another useful fact is that for any polygon, the sum of the exterior angles is 360°. So to find the measure of m below, first find the total of the given exterior angles, and then subtract it from 360°: $360 - (30 + 90 + 90 + 30) = 120°$.

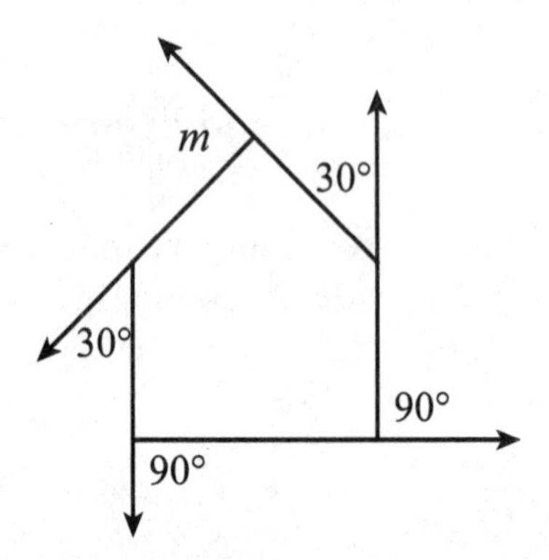

How to Determine the Interior and Exterior Angles of a Quadrilateral

The quadrilateral is a common figure in geometry. The sum of the interior angles is 360°, as is the sum of the exterior angles. Solve these questions as explained in the previous section. There is, however, a special case for any parallelogram; because opposite angles are congruent, any two consecutive angles will sum to 180°. So if you are given a parallelogram with one angle with a measure of 80°, the opposite angle will also be 80°, and the two consecutive angles will be 180° – 80° = 100°. Often you will be faced with an algebraic context with an angle problem. For example, find the value of x in the isosceles trapezoid figure below.

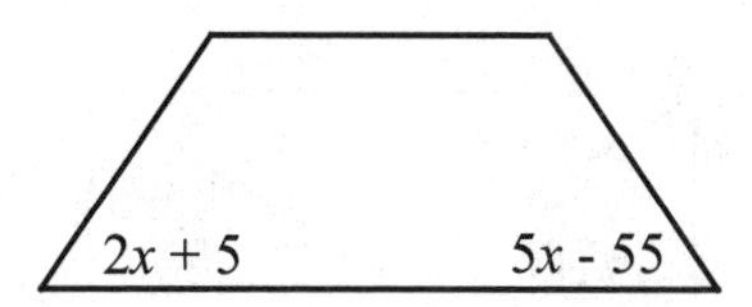

You need to know the fact that these base angles are congruent and set the expressions to be equal.

Now you have an equation to solve:	$2x + 5 = 5x - 55$.
The first step is to subtract $2x$ from both sides:	$2x - 2x + 5 = 5x - 55 - 2x$
	$5 = 3x - 55$
Now, add 55 to both sides:	$55 + 5 = 3x - 55 + 55$
	$60 = 3x$
To find the value of x, divide each side by 3:	$\frac{60}{3} = \frac{3x}{3}$
	$x = 20$

Be sure to read these problems carefully. In this question, you were asked for the value of x. In another instance, you may need to find the measure of the missing angle. In that case, you would substitute the value of the variable x into one of the expressions, like $2x + 5$ to get $2(20) + 5 = 45°$.

Another problem type may give you the ratio of angles as in this question. The consecutive angles in a parallelogram are in the ratio of 2:3. What is the measure of the smaller angle? To solve, recall that the consecutive angles sum to 180°. When presented with a ratio, take each number of the ratio and multiply by x, making an equivalent ratio of $2x$:$3x$. Now the two algebraic expressions can be added and set equal to 180.

The equation is:	$2x + 3x = 180$.
Combine like terms to get:	$5x = 180$.
Divide both sides by 5:	$\frac{5x}{5} = \frac{180}{5}$
	$x = 36°$

The smaller measure is $2x$, or $2(36) = 72°$.

Working with the Sides and the Diagonals of a Quadrilateral

As covered earlier in this chapter, some quadrilaterals have special properties concerning the sides and the diagonals. For example, what is the length of the diagonal AC in the rectangle below if DE = 17 cm?

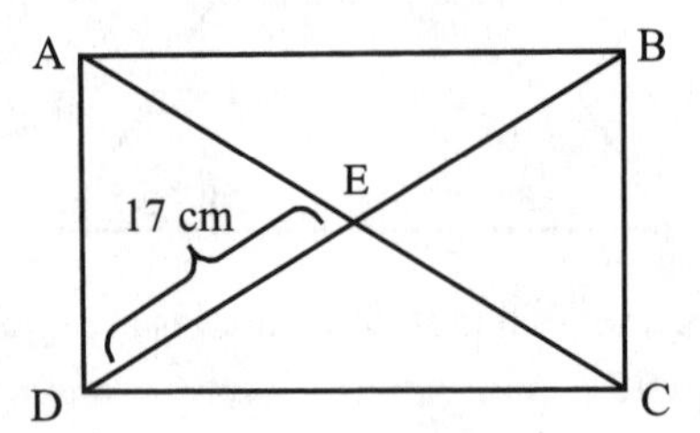

Knowing that the diagonals bisect each other, you can conclude that EB = 17 cm. The length of diagonal BD = 17 + 17 = 34 cm. You also know that the diagonals of a rectangle are congruent, so the length of diagonal AC is also 34 cm.

You may also need to solve an algebraic equation to find a missing length in a quadrilateral. Suppose you are asked to find the length of side EF in the parallelogram below.

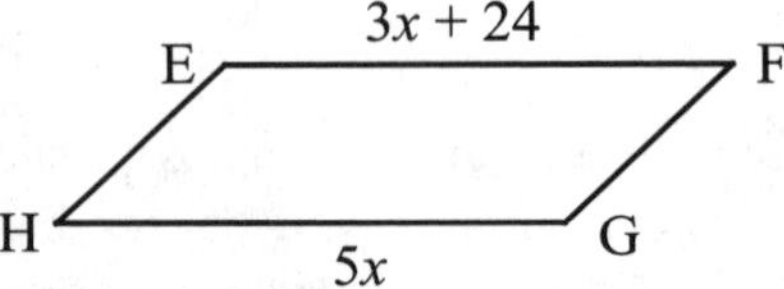

The opposite sides are congruent, so:

Set up the equation: $3x + 24 = 5x$

Subtract $3x$ from both sides: $3x + 24 - 3x = 5x - 3x$

$24 = 2x$

Divide both sides by 2: $\frac{24}{2} = \frac{2x}{2}$

$x = 12$

The expression for side EF is $3x + 24$, so the length is $3(12) + 24$, or $36 + 24 = 60$ units in length.

STEP-BY-STEP ILLUSTRATION OF THE FIVE MOST COMMON QUESTION TYPES

This next section will take you through the five most common question types involving polygons and quadrilaterals.

Question 1: Classifying a Polygon

A four-sided polygon is classified as a(n):

(A) quadrilateral

(B) pentagon

(C) hexagon

(D) octagon

(E) decagon

Polygons are classified according to the number of sides. Become familiar with the polygon names as described in this chapter. **The correct answer is choice (A),** quadrilateral, because it is a four-sided figure. A five-sided polygon is a pentagon, choice (B). A hexagon, choice (C), has six sides. An eight-sided figure is an octagon, choice (D). Choice (E), a decagon, has 10 sides.

Question 2: Finding the Missing Angle in a Regular Polygon

What is the degree measure of an angle in a regular octagon?

(A) 80°

(B) 135°

(C) 180°

(D) 1080°

(E) 1440°

A regular octagon has eight congruent angles. The sum of all the interior angles of a polygon is $180(n - 2)$ where $n = 8$, the number of sides. Therefore, the sum of all the angles is $180(6) = 1080°$. To find the measure of one of the angles, divide the value by 8 to get $1080 \div 8 = 135°$, which is **choice (B), the correct answer.** If your choice was (A), you may have mistakenly thought that the angle would be 10 times the number of sides. If you chose choice (C), you may have multiplied 180(8) instead of 180(8 – 2), and then divided by the 8 sides of an octagon. Choice (E) is the sum of the interior angles in a decagon. Choice (D) is the sum of all the angles in an octagon, not the measure of one of the angles.

Question 3: Finding the Missing Side of a Quadrilateral

Find the measure of side ZY in the parallelogram below:

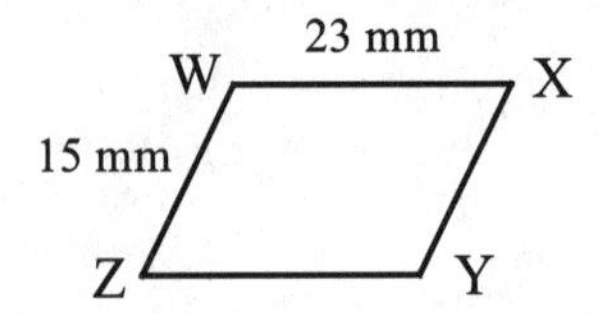

(A) 8 mm

(B) 15 mm

(C) 23 mm

(D) 38 mm

(E) 76 mm

To answer this question, remember that the opposite sides of a parallelogram are congruent. The side opposite to side ZY is side WX, which has a length of 23 mm. Therefore, **the correct answer is choice (C).** Choice (A) is the difference between the two given measures in the diagram. Choice (B) is the length of side XY. Choice (D) is the sum of the two given measures. Choice (E) is the perimeter of the parallelogram.

Question 4: The Diagonals of a Quadrilateral

In the rectangle below, the measure of RT is represented as $6x - 28$, and the measure of QS is represented as $2x + 1$. What is the length of diagonal RT?

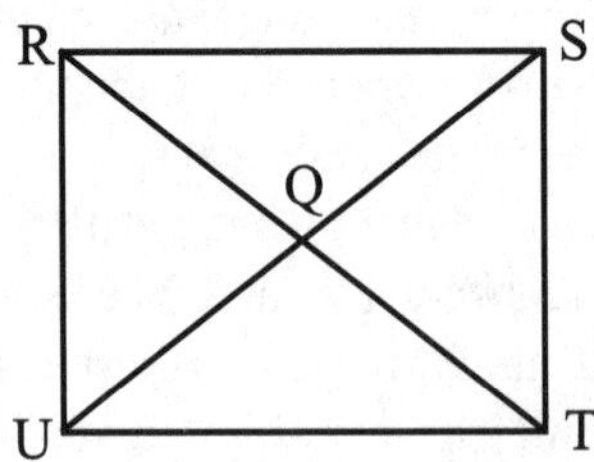

(A) 7 units

(B) 14 units

(C) 15 units

(D) 31 units

(E) 62 units

Remember that the diagonals of a rectangle bisect each other and they are congruent. You will use both of these facts to solve this common problem type. Because the diagonals bisect each other, they divide each other in half. So when you are told that QS is represented as $2x + 1$, the length of UQ is also $2x + 1$. Use the knowledge that the diagonals are congruent to set up the equation:

The equation is:	$2x + 1 + 2x + 1 = 6x - 28$
Combine like terms on the left-hand side:	$4x + 2 = 6x - 28$
Subtract $4x$ from both sides:	$4x + 2 - 4x = 6x - 28 - 4x$
	$2 = 2x - 28$
Add 28 to both sides:	$2 + 28 = 2x - 28 + 28$
Divide both sides by 2:	$\frac{30}{2} = \frac{2x}{2}$
	$15 = x$

Be careful at this point—you are not done yet! Fifteen is the value of x, not the length of RT. This is a common trap you will find on many standardized tests. Substitute in the value of 15 into the expression for RT to find its length: $6x - 28 = 6(15) - 28 = 62$ units. **Choice (E) is the correct answer.** If your answer was choice (A) or (B), you may have mistakenly solved the equation $2x + 1 = 6x - 28$. Choice (A) would be the rounded value of x and choice (B) the incorrect length of RT. Choice (C) is the correct value of x, not the length of RT. Choice (D) is the length of segment QS.

Question 5: Using Ratios with Polygons

The angles in a rhombus are in the ratio of 3:5. What is the measure of the larger angle?

(A) 22.5°

(B) 45°

(C) 67.5°

(D) 112.5°

(E) 225°

When working with ratios, use algebra to solve the problem. Multiply the two values in the ratio by x to get an equivalent ratio of $3x{:}5x$. Use the fact that the consecutive angles in a parallelogram sum to 180°. A rhombus is a special type of parallelogram, so its consecutive angles will have this trait.

Set up the equation as: $3x + 5x = 180$

Combine like terms on the left-hand side: $8x = 180$

Divide both sides by 8: $\frac{8x}{8} = \frac{180}{8}$

$x = 22.5$

Finish up the problem by finding the measure of the larger angle, that is the angle represented by $5x = 5(22.5) = 112.5$, which is **choice (D), the correct answer.** Choosing (A) is a common error made by test takers. This is simply the value of the variable *x*, not the measure of the larger angle. If you multiplied 22.5 by 2 instead of by 5, you would arrive at the incorrect answer (B). You may have also concluded that choice (B) was correct if you set up the equation as $3x + 5x = 360$, and then gave the answer as the value of *x*.

Be sure to read the problem correctly and know what the question is asking you to find. If you did not read carefully, you may have arrived at choice (C), in which case you gave the measure of the smaller angle. If your answer was choice (E), again you set up the equation as $3x + 5x = 360$ and solved for the larger angle based on this misconception.

CHAPTER QUIZ

1. How many sides does a pentagon have?

   (A) 3

   (B) 4

   (C) 5

   (D) 6

   (E) 8

2. What is the degree measure of the angle marked as *x* in the parallelogram below?

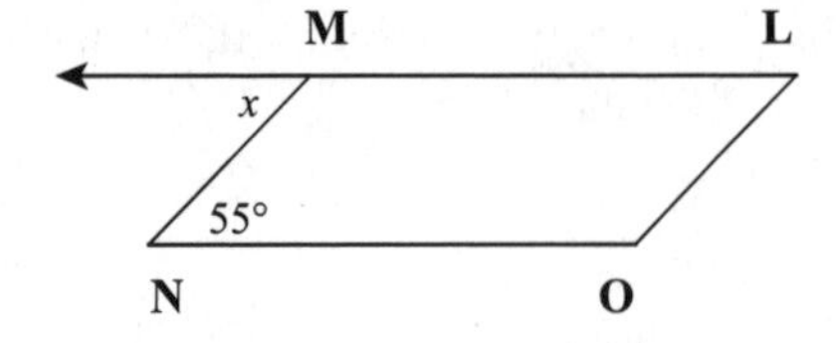

   (A) 25°

   (B) 35°

   (C) 45°

   (D) 55°

   (E) 125°

3. What is the degree measure of an angle in a regular hexagon?
   (A) 60°
   (B) 90°
   (C) 108°
   (D) 120°
   (E) 135°

4. If the sum of the measures of the interior angles of a regular polygon is 540°, what is the name of the polygon?
   (A) Triangle
   (B) Quadrilateral
   (C) Pentagon
   (D) Hexagon
   (E) Octagon

5. What is the length of side SU in the rectangle below?

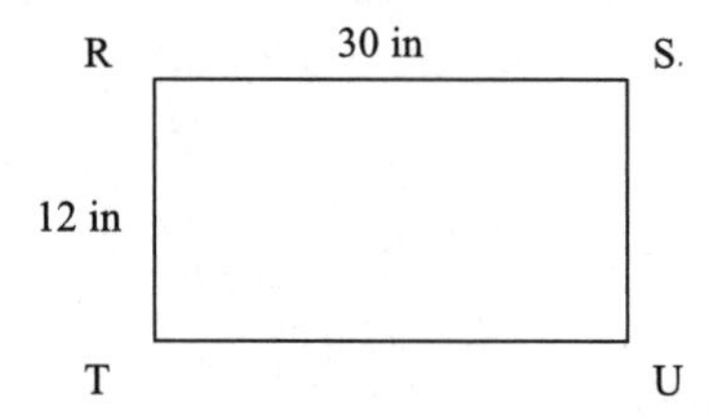

   (A) 12 in
   (B) 18 in
   (C) 24 in
   (D) 30 in
   (E) 42 in

6. Find the value of x in the isosceles trapezoid shown here:

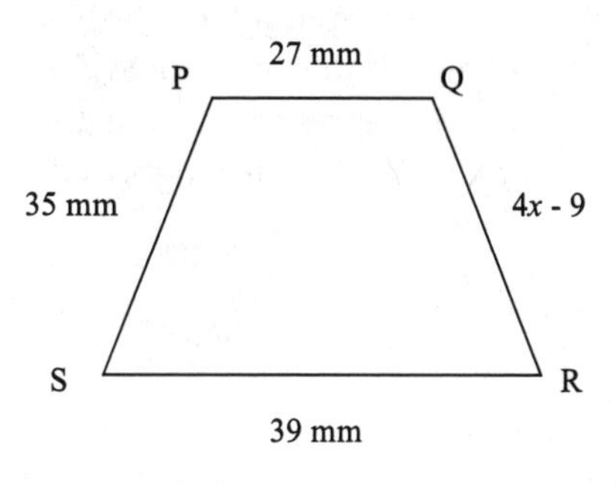

   (A) 4.5 mm
   (B) 6.5 mm
   (C) 9 mm
   (D) 11 mm
   (E) 35 mm

7. In square ABCD below, diagonal AC = 28 cm. What is the length of segment DE?

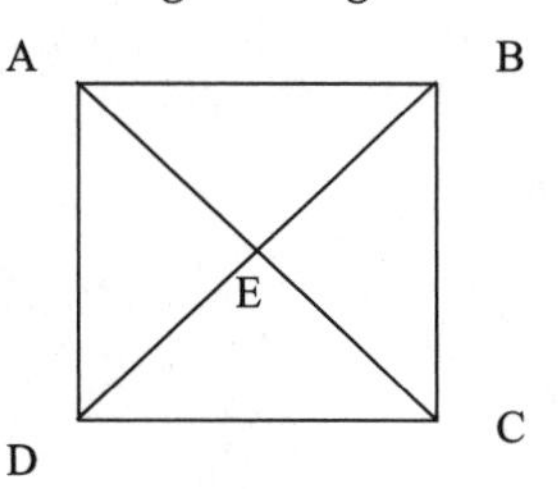

   (A) 7 cm
   (B) 10 cm
   (C) 14 cm
   (D) 28 cm
   (E) 56 cm

8. Parallelogram WXZY is shown below, with diagonals XY and WZ. The expression $7x + 4$ represents the length of WZ and $6x + 16$ represents the length of XY. What is the length of the diagonal WZ?

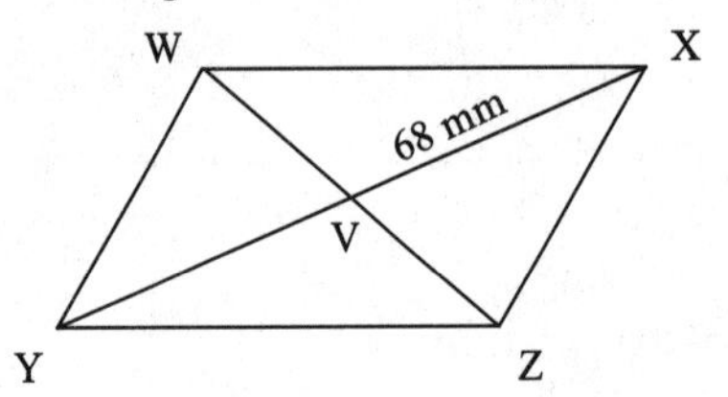

(A) 33 mm

(B) 68 mm

(C) 72 mm

(D) 136 mm

(E) 144 mm

9. The angles in a rhombus are in the ratio of 4:5. What is the measure of the smaller angle?

(A) 20°

(B) 80°

(C) 90°

(D) 100°

(E) 180°

10. The sides of a rectangle are in the ratio of 3:7. If the longer side is 630 m, what is the length of the shorter side?

(A) 63 m

(B) 90 m

(C) 210 m

(D) 270 m

(E) 1470 m

ANSWER EXPLANATIONS

1. C

Polygons are classified according to the number of sides they have. A pentagon has five sides. A three-sided polygon is a triangle. A four-sided polygon is a quadrilateral. A six-sided polygon is called a hexagon; an eight-sided polygon is called an octagon.

2. D

Recall your facts about parallelograms and exterior angles. First, find the measure of the interior angle $\angle LMN$. In a parallelogram, the adjacent angles sum to 180°. Therefore the measure of$\angle LMN + 55° = 180°$, or $\angle LMN = 180° - 55° = 125°$. Now that you have the value of $\angle LMN$, recall that the exterior angle and the adjacent interior angle are a linear pair; their measures sum to 180° as well. So the measure of the exterior angle, x, is $180 - 125 = 55°$.

Choice (B) represents the complement to the angle of 55°. If your answer was choice (E), you found the measure of $\angle LMN$, instead of the exterior angle.

3. D

If n represents the number of sides in a polygon, then the sum of the measures of the interior angles of the polygon is $180\ (n - 2)$. A hexagon has six sides, so the sum of all of the interior angles is $180\ (6 - 2) = 180(4) = 720°$. This is the sum of the angles. To find the measure of any one of the angles, divide this sum by 6, because there are six congruent angles in a regular hexagon. So $720° \div 6 = 120°$.

Choice (A) is the measure of an angle in an equilateral triangle. Choice (B) is the measure of an angle in a square. The measure of 108° would be the measure of one angle in a regular pentagon. Choice (E), 135°, would be the interior angle measure of a regular octagon. If your choice was (E), you may have forgotten to subtract 2 from the number of sides in the hexagon.

4. C

For this problem, you can use the formula $S = 180\ (n - 2)$, where S represents the sum of the interior angles in a regular polygon and n represents the number of sides in the regular polygon. Substituting into the formula, you get:

The equation is:	$540 = 180(n - 2)$
Divide both sides of this equation by 180:	$\frac{540}{180} = \frac{180(n-2)}{180}$
Simplify:	$3 = n - 2$
Add 2 to both sides to isolate n:	$3 + 2 = n - 2 + 2$
	$n = 5$

A five-sided polygon is a pentagon.

5. A

In a rectangle, as is true for any parallelogram, opposite sides are congruent. The side opposite SU is RT with a length of 12 inches, choice (A). Choice (B) is the difference between length RS and RT. Choice (D) is the length of side TU. Choice (E) is the sum of the two given lengths.

6. D

In an isosceles trapezoid, the nonparallel sides are congruent. These are sides PS and QR. Set these expressions equal to each other and solve for x.

The equation is:	$4x - 9 = 35$
Add 9 to both sides of the equation:	$4x - 9 + 9 = 35 + 9$
Divide both sides by 4:	$\frac{4x}{4} = \frac{44}{4}$
	$x = 11$

Be careful; you are not asked to find the value of the expression $4x - 9$, just the value of x. If your answer was choice (A), you may have set up an incorrect equation as $4x - 9 = 27$ and then made a second mistake by subtracting 9 from each side. Choice (B) would be the result of the correct equation $4x - 9 = 35$, but then subtracting 9 from each side instead of adding 9. Choice (C) would be the result of solving for x with the incorrect equation $4x - 9 = 27$. Choice (E) is the length of QR, not the value of x.

7. C

A square is both a rectangle and a parallelogram. In this type of quadrilateral, the diagonals bisect each other and are congruent. Because they are congruent, you can conclude that diagonal BD is 28 cm. Because the diagonals bisect one another, DE is one half of 28, or 14 cm, choice (C). You may have chosen (A) if you were careless and thought that the length was one-fourth of the length of the diagonal. If your choice was (E), you may have mistakenly multiplied by 2 instead of dividing.

8. E

In a parallelogram, the diagonals bisect each other. They are not necessarily congruent—this would only be true for a rectangle or a square. From the diagram, you see that VX = 68 mm. Because the diagonal is bisected, the length of XY = 68 • 2 = 136 mm. Set this equal to the expression $6x + 16$ and solve for x:

The equation is:	$6x + 16 = 136$
Subtract 16 from each side of the equation:	$6x + 16 - 16 = 136 - 16$
Divide both sides by 6:	$\frac{6x}{6} = \frac{120}{6}$
	$x = 20$

Now that you have the value of x, substitute the 20 into the expression of WZ, $7x + 4$ to get $7(20) + 4 = 140 + 4 = 144$ mm. Be careful to keep your quadrilateral facts straight. The values of 68, choice (B) and 136, choice (D), represent the part or the whole of diagonal XY, which is not congruent to diagonal WZ. Choice (C) is the length of one half of the diagonal of WZ, which is WV or VZ.

9. B

You are told that the angle measures are in the ratio of 4:5. Multiply each piece of this ratio by x to get the equivalent ratio $4x{:}5x$. Because a rhombus is a parallelogram, consecutive angles sum to 180°.

Set up an equation as:	$4x + 5x = 180$
Combine like terms:	$9x = 180$
Divide both sides by 9:	$\frac{9x}{9} = \frac{180}{9}$
	$x = 20$

You are not done yet; you are asked for the measure of the smaller angle. The smaller angle is represented as $4x$, or $4(20) = 80°$, choice (B). Choice (A) is the value of the variable x. If your choice was (C), you may have mistakenly thought a rhombus was a rectangle. Choice (D) is the measure of the larger of the two angles, and choice (E) is the sum of the two angles.

10. D

If the sides are in the ratio of 3:7, you can multiply each part of the ratio by x to get an equivalent ratio of $3x{:}7x$. You are told that the longer side, $7x$, is 630 m in length,

So set up the equation: $7x = 630$

Divide each side of this equation by 7: $\frac{7x}{7} = \frac{630}{7}$

$x = 90$

The shorter side is 3 (90) = 270 m, choice (D). If your answer was choice (A), you incorrectly set up an equation stating that $3x + 7x = 630$, and then solved for x. Choice (B) represents the correct value of x, not the length of the shorter side. If you chose answer choice (C) or (E), you made the mistake of thinking the shorter side was 630 m and set up the incorrect equation $3x = 630$. This would result in $x = 210$, choice (C). Choice (E) would be the length of the longer side using this incorrect value of 210 for x.

CHAPTER 4

Circles

WHAT ARE CIRCLES?

A circle is another geometric figure. A *circle* is the collection of points that are all the same distance from a given point, called the *center*. The center point defines the circle, but it is not a point on the circle. This chapter will identify many elements and geometric figures related to the circle. The next section will go on to explore many facts concerning this important concept in geometry.

CONCEPTS TO HELP YOU

The concepts section will cover the basic information needed to solve circle-related problems. First, you will learn about chords, the special chord called the diameter, and the radius. Angles and arcs are additional elements associated with circles. This section will continue by identifying the two types of units used to measure angles and arcs. We'll conclude this section with two special segments related to circles—the secant and the tangent.

Chords of a Circle

A *chord* of a circle is any line segment whose endpoints are on the circle. One special chord is the *diameter*, which passes through the center of the circle. One half of a diameter is called a *radius*; the radius is a segment whose endpoints are the center of the circle and any point on the circle. The figure below shows some examples of chords, including the diameter. A radius is also shown.

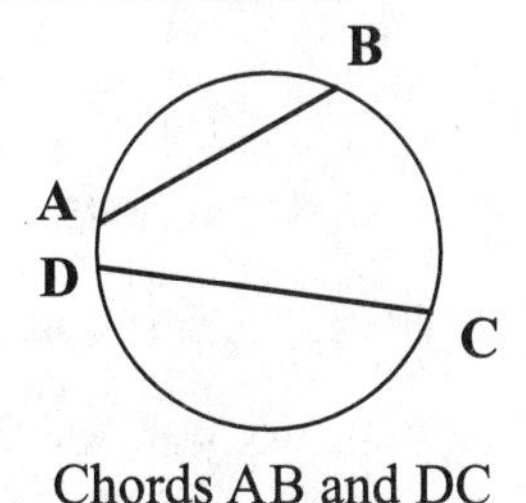

Chords AB and DC

Circle O with diameter LN and radius OM shown

The material in this chapter will deal with what is called the unit circle, whose radius is considered to be 1 unit in length.

Angles and Arcs of a Circle

There are two important angles related to the circle—central angles and inscribed angles. A *central angle* is an angle whose vertex is at the center of the circle. An *inscribed angle* is one in which the vertex is on the circle itself.

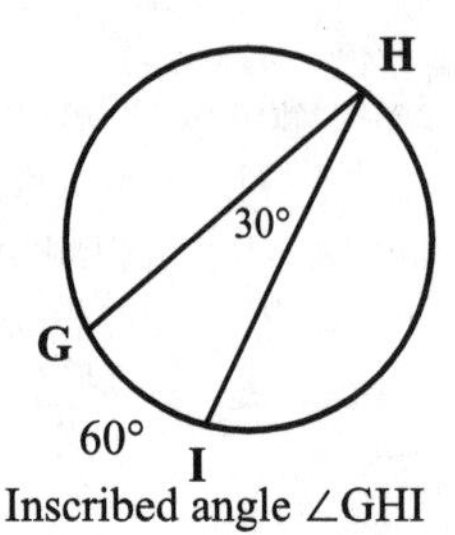

Inscribed angle ∠GHI

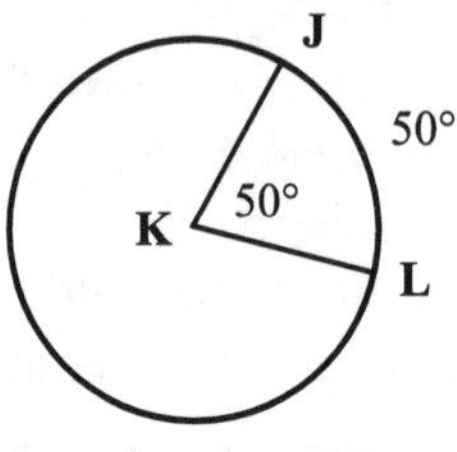

Central angle ∠JKL

These angles can be measured in degrees, as shown above. There is a total of 360° in a circle. An *arc* is simply a piece of the circle. An arc has a definite measure, which is related to the angle measures of either a central or inscribed angle. The arc that is within the interior of a central angle has a measure equal to the measure of the central angle. The arc that is contained within the interior of an inscribed angle is twice the measure of the angle's measure. These measures are shown in the figure above.

Some important concepts deal with the chords and arcs of a circle.

Chords and Arcs

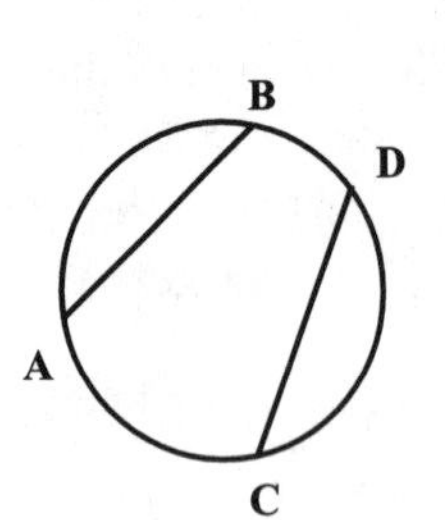

Figure I
If chords AB = CD, then arc AB = arc CD

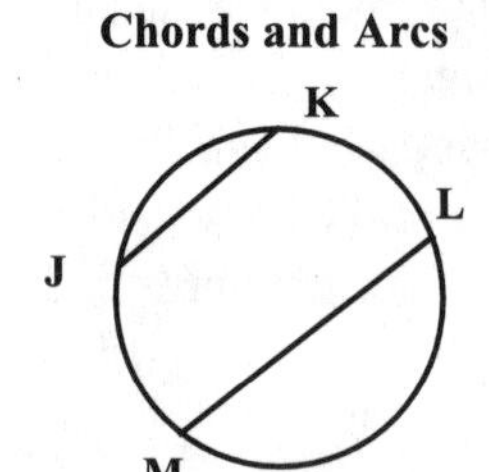

Figure II
If chords JK and LM are parallel, then arc JM = arc KL

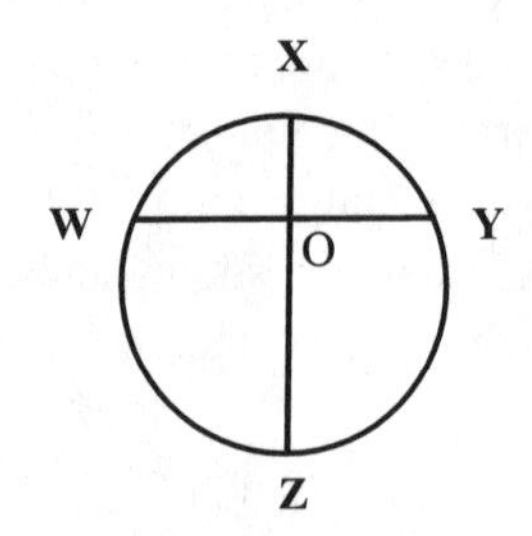

Figure III
If diameter XZ is perpendicular to chord WY, then WO = OY, arc WX = arc XY, and arc WZ = arc YZ

Another fact relates the measure of intersecting chords with the angles created by the intersection, as shown below.

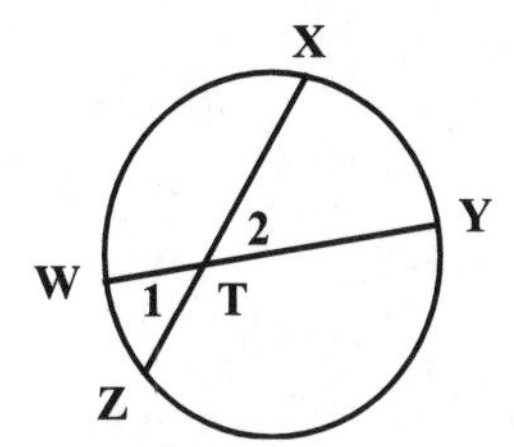

Recall from Chapter 1 that angles 1 and 2 above are vertical angles and that they have the same measure. Either of these angles is related to the intercepted arcs in that the angle measure is one-half of the sum of the intercepted arcs:
$m\angle WTZ = m\angle XTY = \frac{1}{2}$(measure of arc XY + measure of arc WZ).

When two chords intersect in a circle, the chord segments have a special relationship that is proportional. The product of the two segments that make up one of the chords is equal to the product of the two segments that make up the second chord. In the figure above, $\overline{WT} \bullet \overline{TY} = \overline{TZ} \bullet \overline{XT}$.

Radians and Degrees

As mentioned in the previous section, arcs and angles of a circle can be measured in degrees. Another common unit used to measure angles and arcs is *radians*. The relationship between degrees and radians is that 360° = 2π radians. Use this relationship if you need to convert from one unit to another. The symbol π, pronounced *"pi,"* is an irrational number that is the ratio of the *circumference*, the distance around a circle, to its diameter.

Tangents and Secants

There are two types of lines to consider with circles—secants and tangents. A *secant line* is any line that crosses through any two points on the circle. A *secant segment* is a segment whose one endpoint is on the circle and the other endpoint is outside the circle. A secant segment contains the other point on the circle. A *tangent line* is a line that passes through exactly one point on the circle, called the *point of tangency*. A *tangent segment* is a line segment whose one endpoint is the point of tangency and the other endpoint is outside the circle. The figure below illustrates these concepts.

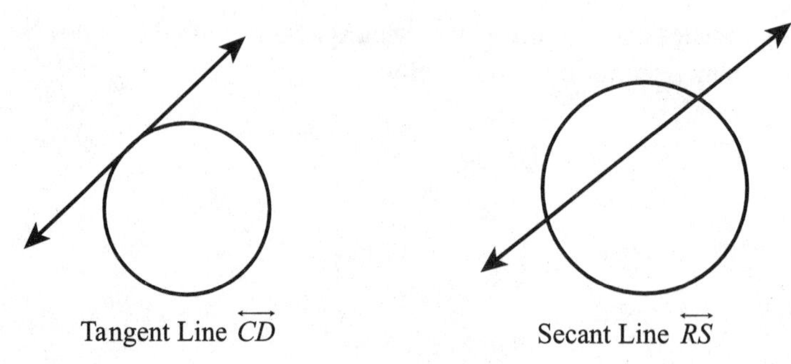

As with chords, there are a number of theorems defining the measures of arcs, angles, and the secant and tangent segments of a circle.

Tangents, Secants and Exterior Angles

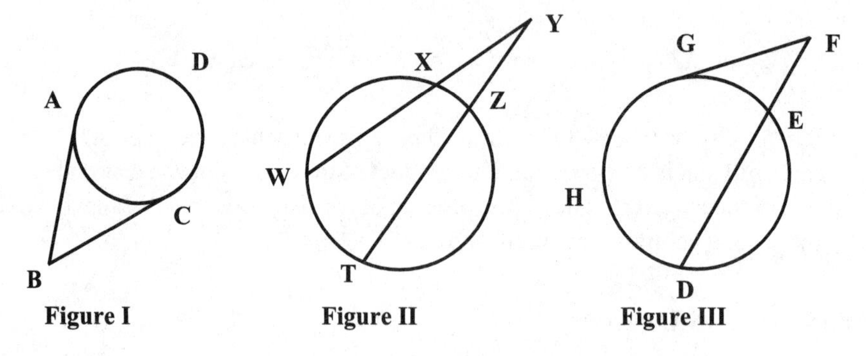

TANGENTS, SECANTS, AND ANGLES

The angle formed by two tangents, ∠ABC in Figure I above, is equal to one half of the difference between the intercepted arcs:

$m\angle ABC = \frac{1}{2}$(measure of arc ADC - measure of arc AC)

This is true for any of the angles above, whether formed by two tangents (Figure I), two secants (Figure II), or a tangent and a secant (Figure III). So $m\angle WYT = \frac{1}{2}$(measure of arc WT - measure of arc XZ) in (Figure II) and $m\angle DFG = \frac{1}{2}$(measure of arc DHG - measure of arc EG)

Two tangent segments to a circle that share the exterior endpoint are equal in length. Thus, in Figure I above, $\overline{AB} = \overline{CB}$.

The measure of secant segments that share the exterior endpoint follows the relationship that the product of one outer secant segment and the whole secant segment is equal to the product of the other outer secant segment and the whole secant segment. Therefore, in Figure II above, $\overline{XY} \bullet \overline{WY} = \overline{ZY} \bullet \overline{TY}$.

When there is a tangent and a secant that share the exterior endpoint, the tangent segment length squared is equal to the product of the outer secant segment and the whole secant segment. So for Figure III above, $(\overline{GF})^2 = \overline{EF} \bullet \overline{DF}$.

STEPS YOU NEED TO REMEMBER

This section will cover the steps that you need to perform to solve problems dealing with the circle. You will use all of the theorems explained in the concepts section.

Finding an Angle Measure When the Intercepted Arc Measure Is Known

If you know the measure of an arc, you can find the measure of an inscribed or central angle. To find an inscribed angle measure, divide the arc measure by 2. For example, the figure below shows inscribed angle ∠JKL, with intercepted arc whose measure is 100°. The measure of angle ∠JKL is 100 ÷ 2 = 50°.

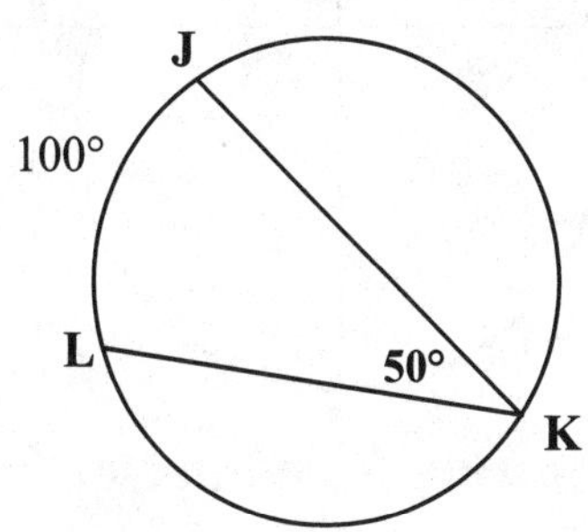

If you are given the measure of an arc that is intercepted by a central angle, the central angle measure will equal the arc measure. This was explained earlier in the chapter in the Angles and Arcs of a Circle section.

Converting from Degrees to Radians

When working with circles, you may be given problems that measure arcs and angles with either degree units or radian units. The relationship between degrees and radians is that 360° = 2π radians. Radian measures are left in terms of the irrational constant π. To convert from one type of unit to the other, you can use the proportion

$$\frac{\text{deg}}{radians} = \frac{360}{2\pi}$$

For example, to convert 90° to radian measure:

Set up the proportion: $\frac{90}{x} = \frac{360}{2\pi}$

Cross-multiply to get: $90 \cdot 2\pi = 360 \cdot x$

Multiply on the left-hand side to get: $180\pi = 360x$

Divide both sides by 360: $\frac{180\pi}{360} = \frac{360x}{360}$

$\frac{\pi}{2} = x$

So 90° is $\frac{\pi}{2}$ radians.

Determining the Length of Chords in a Circle

If a diameter is perpendicular to a chord, recall that the diameter bisects the chord. Therefore, because AC is 8 cm, CB is also 8 cm and the measure of chord AB is 8 + 8 = 16 cm.

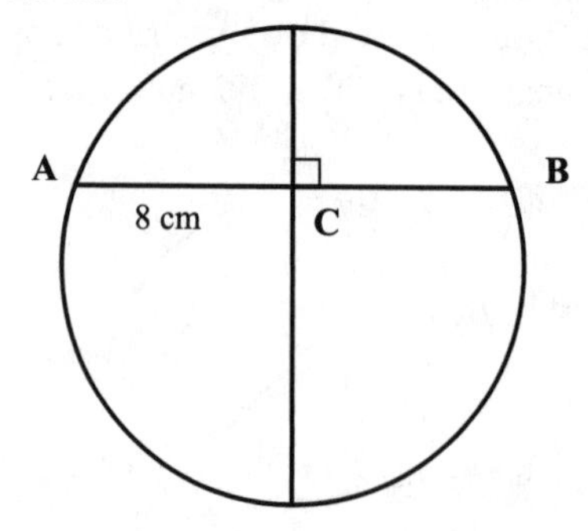

Finding a Missing Measure from Intersecting Chords in a Circle

When two chords intersect in a circle, the segment pieces of the chords are in a proportional relationship. Look at the circle below with intersecting chords DF and EG.

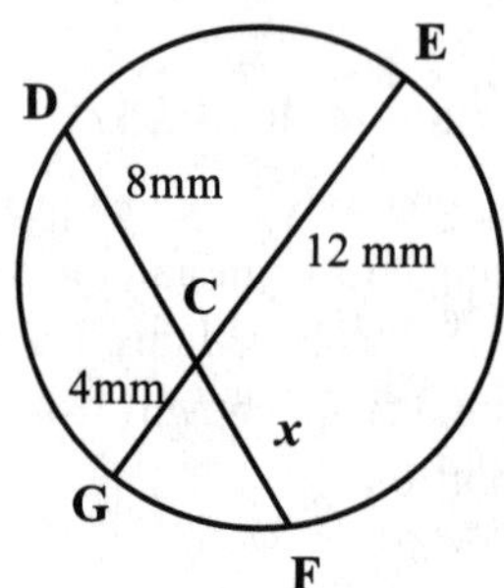

The chord pieces are in the relationship. To find the missing measure of segment piece CF, let x represent the measure of CF, substitute in the given values, and solve the equation for x.

Solve the equation: $12 \bullet 4 = 8x$

Multiply on the left hand side: $48 = 8x$

Divide both sides by 8: $\frac{8}{48} = \frac{8x}{8}$

$6 = x$

So the measure of CF = 6 mm.

Finding a Missing Angle Measure Formed by Two Secants to a Circle

When two secants meet in the exterior of a circle, an angle is formed. This angle intercepts the circle in two places. The figure below shows secant segments AC and EC.

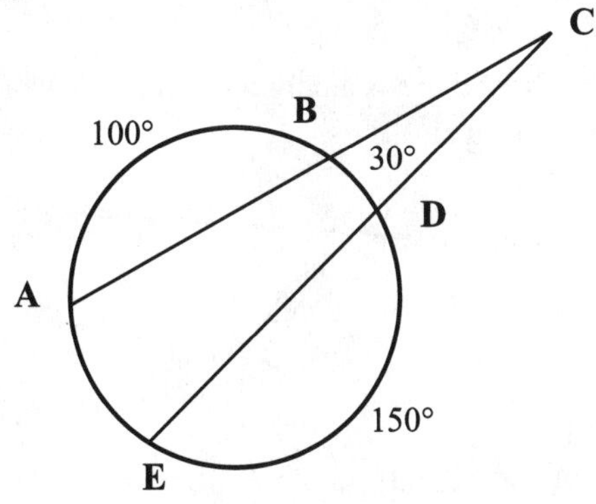

The arc measures are as shown; to find the measure of angle $\angle ACE$, use the theorem that $m\angle ACE = \frac{1}{2}$ (measure of arc AE – measure of arc BD). The measure of arc AE is not given; however, recall that the entire circle has measure of 360°, so the measure of arc AE is 360 – (100 + 30 + 150) = 360 – 280 = 80°. Now substitute in the values of arcs AE and BD to find the measure of angle $m\angle ACE = \frac{1}{2}(80 - 30) = \frac{1}{2} \times 50 = 25°$.

Finding a Missing Tangent Measure When There Are Two Tangents to a Circle

Any two tangent segments to a circle that meet have the same measure. Just remember this fact to find a missing tangent length. In the circle below, tangents LM and MN are congruent; segment MN has the same measure as segment LM, of 15 cm.

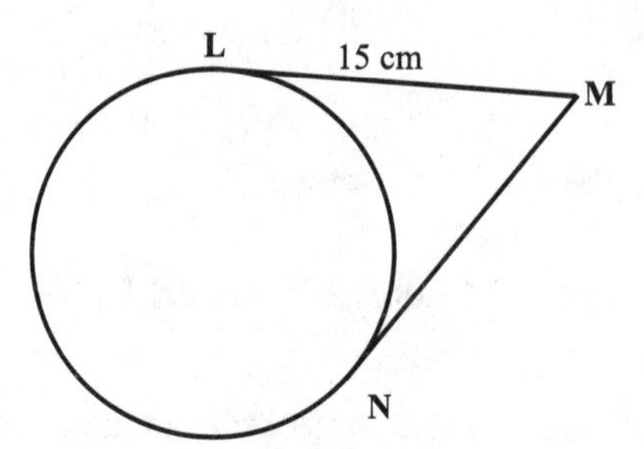

Find a Missing Tangent Measure When There Is a Tangent and a Secant to a Circle

The final case is where there is a tangent segment and a secant segment that meet in the exterior of a circle.

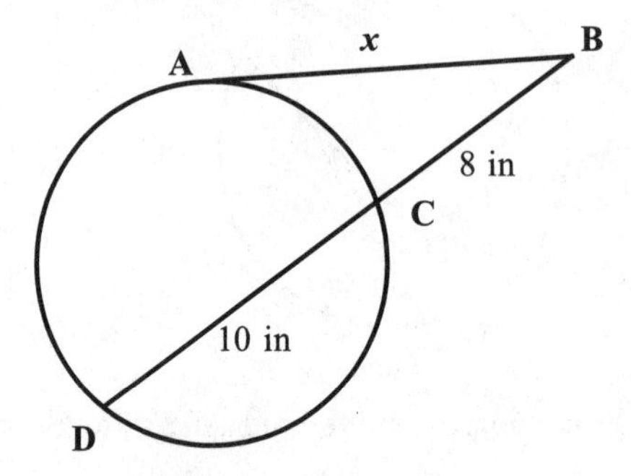

The square of the length of tangent AB is equal to the product of the outer secant segment (CB) and the whole secant segment (DB). The length of secant segment DB is $10 + 8 = 18$ inches. The equation will be $AB^2 = CB \cdot DB$.

Set up the equation:	$x_2 = 8 \cdot 18$
Multiply on the right hand side:	$x_2 = 144$
Take the square root of each side:	$\sqrt{x^2} = \sqrt{144}$
	$x = 12$

The length of tangent $AB = 12$ inches.

STEP-BY-STEP ILLUSTRATION OF THE 5 MOST COMMON QUESTION TYPES

Now, let's take a look at five of the most prevalent problem types you may encounter in geometry problems related to circles, segments, and angles. After completing this section, in conjunction with the previous section, you will have worked through the various ways that these types of problems are presented.

Question 1: Finding the Measure of an Arc from the Measure of an Inscribed Angle

What is the measure of arc YZ in the circle below?

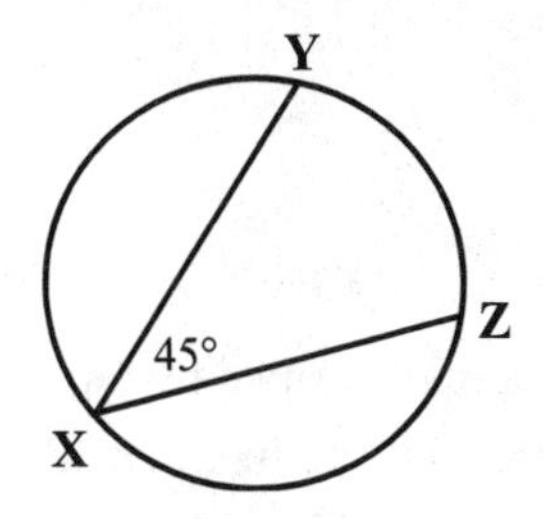

(A) 11.25°

(B) 22.5°

(C) 45°

(D) 90°

(E) 180°

In the figure, angle ∠YXZ is an inscribed angle; its   is on the circle itself. For an inscribed angle, the measure of the arc that is intercepted by the angle is twice the measure of the angle. The angle measures 45°, so the measure of arc YZ is **the correct answer choice (D)**, 90°.

Choice (A) is one-fourth of the measure of angle ∠YXZ. If your choice was (B), 22.5°, you calculated one-half of the angle measure, instead of twice the measure. Choice (E) is four times the measure of the angle.

Question 2: Finding the Measure of an Angle Formed From Two Intersecting Chords

What is the measure of angle ∠AED in the circle below?

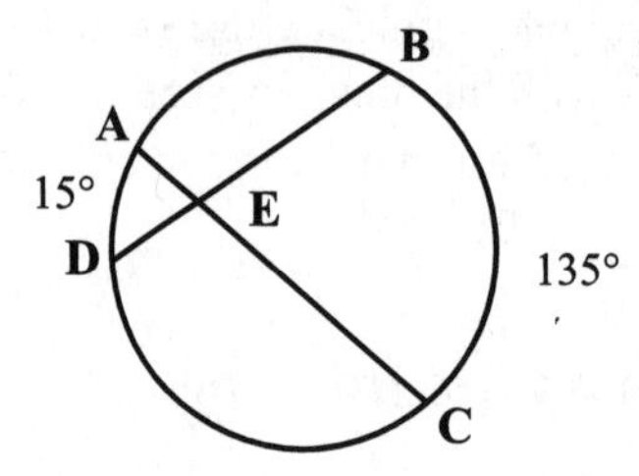

(A) 15°

(B) 60°

(C) 75°

(D) 135°

(E) 150°

When two chords intersect in a circle, the interior angles formed by this intersection have a measure that is related to the arcs that are intercepted. In this figure, ∠AED and ∠BEC intercept arcs of lengths 15° and 135°. The angle is one-half of the sum of the intercepted arcs. So angle $\angle AED = \frac{1}{2}(15+135) = \frac{1}{2}(150) = 75°$, **the correct answer choice (C).**

Choice (A) is the measure of arc AD. If your choice was (B), you calculated one-half of the difference between the arcs, instead of correctly figuring one-half of the sum of the arcs. Choice (D) is the arc BC. If you forgot to multiply the sum by one-half, you would arrive at choice (E).

Question 3: Finding the Measure of a Secant Segment Given Two Secants

What is the measure of secant XZ in the figure below?

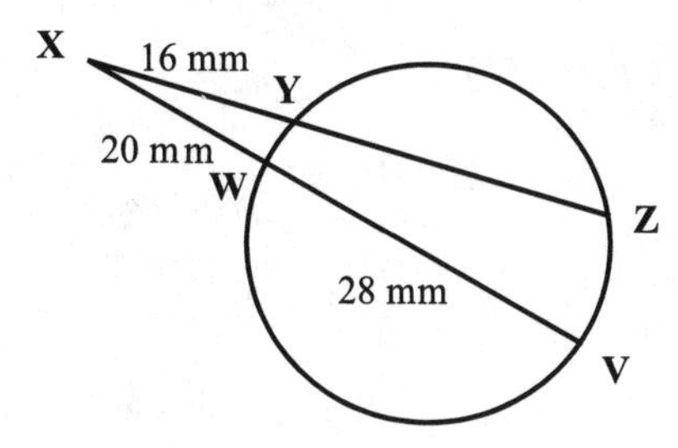

(A) 35 mm

(B) 28 mm

(C) 44 mm

(D) 51 mm

(E) 60 mm

When two secants meet in the exterior of a circle, their segment lengths are in a proportional relationship. The product of the outer segment and the whole secant is equal to the product of the other outer segment and the whole secant. The whole secant segment XV is 20 + 28 mm = 48 mm in length; the outer piece is 20 mm. Segment XY is given as 16 mm, and you are asked to find the length of secant XZ. Use the variable m to represent the length of secant XZ.

Set up the equation: $20 \cdot 48 = 16m$

Multiply on the left: $960 = 16m$

Divide both sides by 16: $\frac{960}{16} = \frac{16m}{16}$

$60 = m$

So the correct answer is choice (E).

If your choice was (A) or (D), you fell into a common mistake when solving these secant length problems. You used the product of the two segments instead of the outer segment and the whole secant. Choice (A) is an incorrect measure for the inner segment YZ. Choice (D) is an incorrect length for secant XZ. Choice (B) is the length of segment WV, and you cannot conclude that these two inner segments have the same length. If your choice was (C), you found the length of segment YZ instead of the length of the whole segment.

Question 4: Finding the Measure of an Angle formed by Two Tangents

If the measure of arc FH is 125° in the circle shown below, what is the measure of angle ∠FGH?

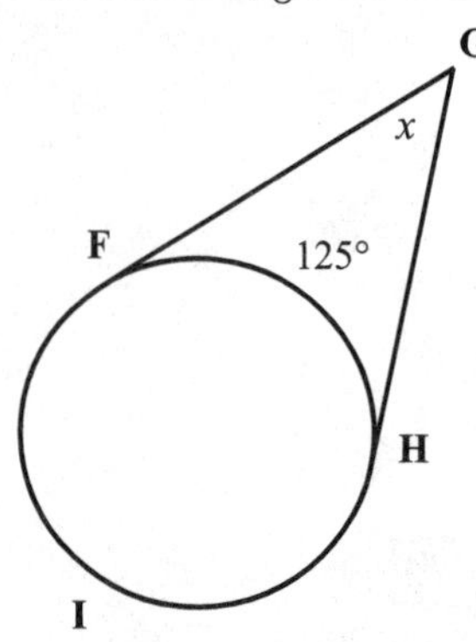

(A) 55°

(B) 62.5°

(C) 117.5°

(D) 180°

(E) 235°

The angle formed by two tangents to a circle is equal to one-half of the difference between the arcs that are intercepted. At first glance, you may think that there is not enough information given in this problem; no measure is given for arc FIH. But recall the fact that the entire circle is 360°, and the two intercepted arcs to tangents sum to the whole circle. So arc FIH = 360° – 125° = 235°. So the angle ∠FGH, represented as *x,* is $\frac{1}{2}(235-125)=\frac{1}{2}(110)=55°$, **the correct answer choice (A).**

Choice (B) is one-half of arc measure FH. Likewise, choice (C) is one-half of arc measure FIH. Choice (D) is a common error, in that you may have calculated the sum of the arcs instead of the difference. If your choice was (E), you found the measure of arc FIH instead of finishing the problem to find the measure of angle ∠FGH.

Question 5: Finding the Length of a Secant Segment Piece Given a Secant and a Tangent

What is the length of chord RS in the circle below?

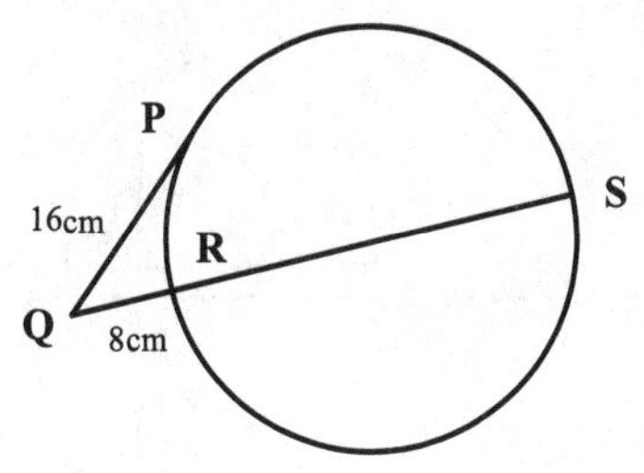

(A) 4 cm

(B) 8 cm

(C) 16 cm

(D) 24 cm

(E) 32 cm

Chord RS is also part of secant QS. This is a secant and a tangent that meet in the exterior of the circle. The tangent length squared is equal to the product of the outer segment of the secant and the length of the whole secant. So let *m* represent the whole secant length QS.

Set up the equation:	$16^2 = 8m$
Square the left hand side:	$256 = 8m$
Divide both sides by 8:	$\frac{256}{8} = \frac{8m}{8}$
	$32 = m$

The length of secant QS is 32 cm. You are asked to find the length of chord RS, which is the whole secant minus the outer piece: $32 - 8 = 24$ cm, which is **choice (D), the correct answer.**

If your answer was choice (A), you may have incorrectly calculated 162 to be $16 \cdot 2$ instead of $16 \cdot 16$. Choice (B) is the length of QR, not the length of RS; choice (C) is the length of QP. Choice (E) could result from a common error. You may have set up the correct equation and solved correctly, but gave the solution to the equation, not the length of the secant segment piece that was asked for.

CHAPTER QUIZ

1. If an inscribed angle in a circle is 42°, what is the measure of its intercepted arc?

   (A) 21°

   (B) 42°

   (C) 48°

   (D) 84°

   (E) 138°

2. Given that O is the center of the circle below, what is the measure of angle ∠AOB?

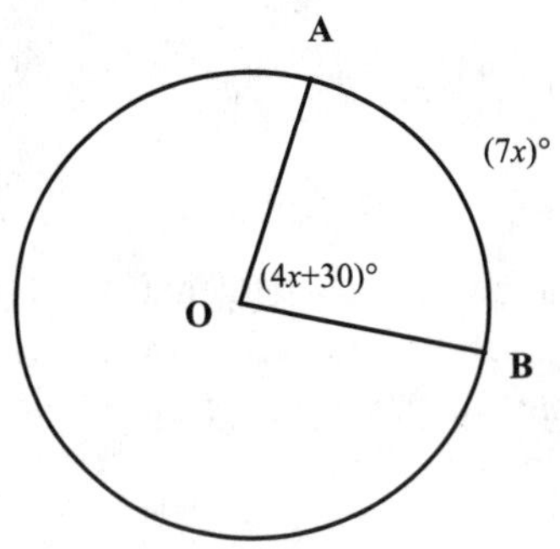

   (A) 10°

   (B) 13.64°

   (C) 20°

   (D) 35°

   (E) 70°

3. What is the radian measure equivalent of 270°?

   (A) $\frac{3}{2}$

   (B) $\frac{3\pi}{4}$

   (C) $\frac{3\pi}{2}$

   (D) 2π

   (E) 3π

4. Given that chord WY is a diameter in the circle below, what is the length of chord XZ?

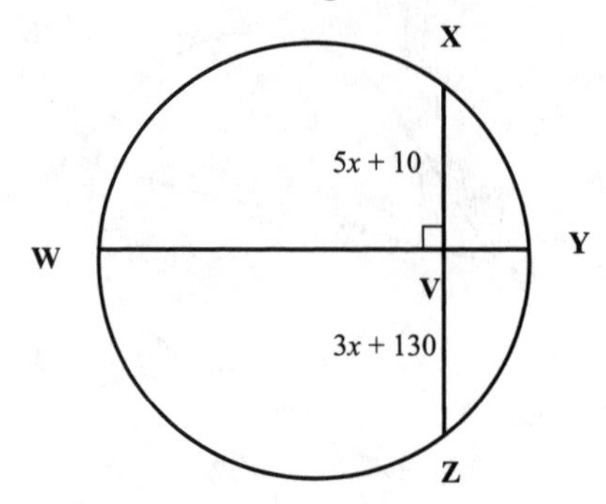

   (A) 5 units

   (B) 60 units

   (C) 140 units

   (D) 310 units

   (E) 620 units

5. What is the length of segment EI in the circle below?

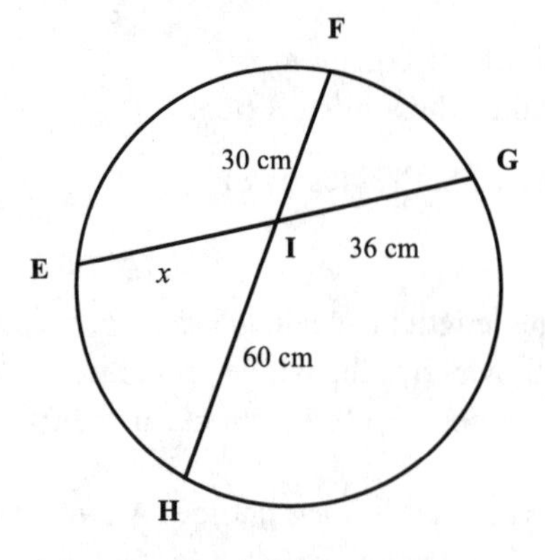

   (A) 50 units

   (B) 72 units

   (C) 86 units

   (D) 90 units

   (E) 126 units

6. What is the measure of arc AD in the following circle?

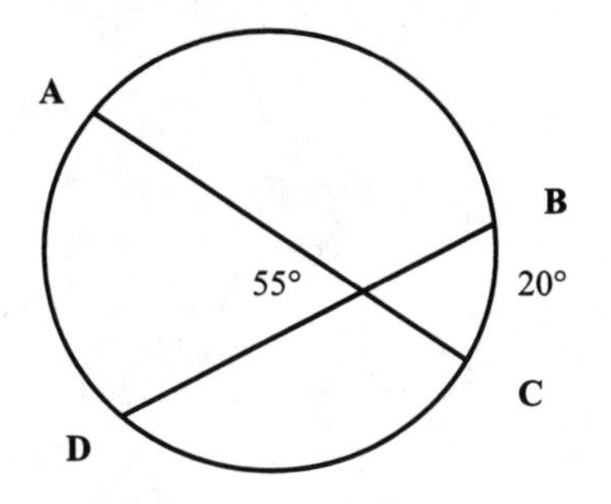

(A) 20°
(B) 55°
(C) 90°
(D) 110°
(E) 130°

7. The circle below shows two tangent segments JK and KL. What is the measure of angle ∠JKL?

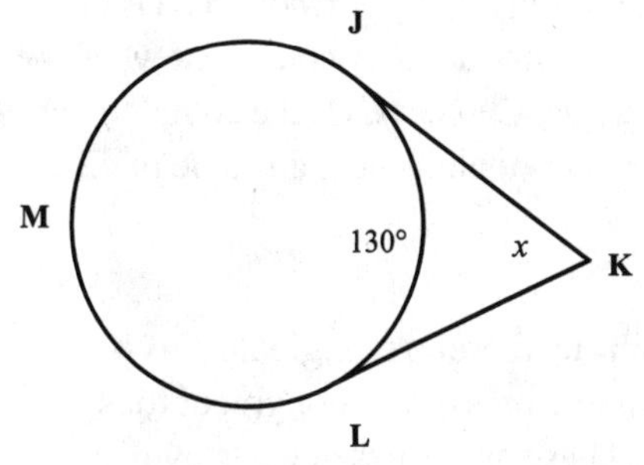

(A) 50°
(B) 65°
(C) 130°
(D) 180°
(E) 230°

8. What is the measure of secant segment QS below?

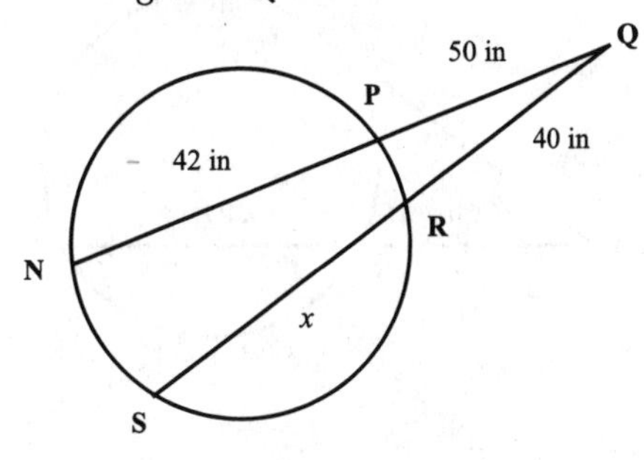

(A) 48 in
(B) 75 in
(C) 115 in
(D) 132 in
(E) 3000 in

9. The following circle shows two secants that form angle ∠TVX. What is the measure of this angle?

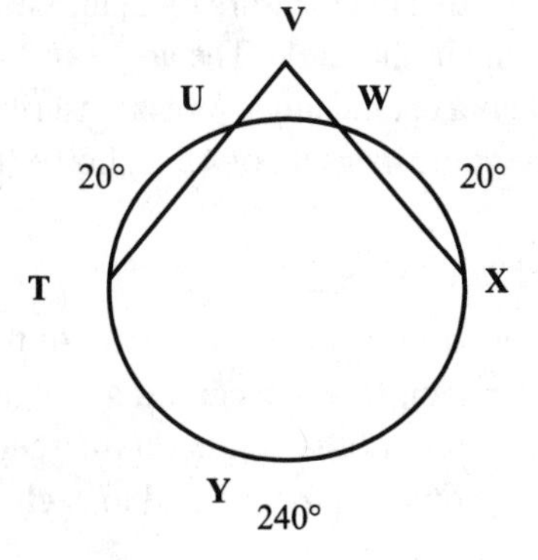

(A) 10°
(B) 20°
(C) 40°
(D) 80°
(E) 160°

10. Given the following circle, what is the measure of tangent segment AB?

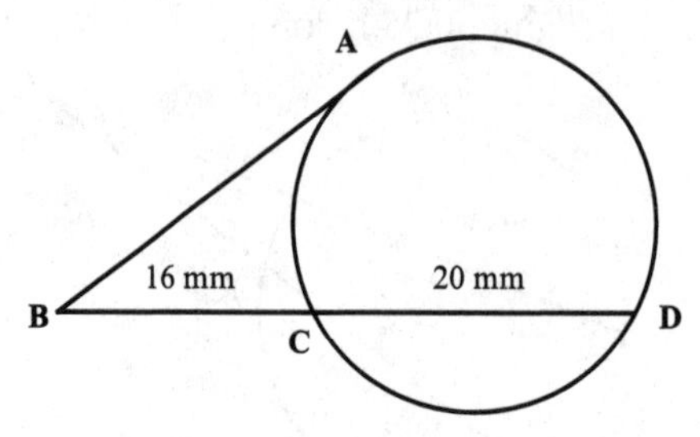

(A) 20 mm

(B) 24mm

(C) 36 mm

(D) 288 mm

(E) 576 mm

ANSWER EXPLANATIONS

1. D

The measure of an arc intercepted by an inscribed angle is two times as large as the measure of the inscribed angle. If your answer was choice (A), you divided the measure by 2, instead of multiplying by 2. Choice (B) is the measure of the angle. The angle and arc measures are only congruent when the angle is a central angle, not an inscribed angle. Choice (C) is the complement to the angle measure of 42°. Choice (E) is the supplement to an angle of 42°.

2. E

You are told that O is the center of the circle; therefore, angle $\angle AOB$ is a central angle. For a central angle, the angle measure is congruent to the measure of the arc that it intercepts. These are represented as two expressions. Set them equal to each other and solve for x:

The equation is: $7x = 4x + 30$

Subtract $4x$ from both sides: $7x - 4x = 4x + 30 - 4x$

Simplify: $3x = 30$

Divide both sides by 3: $\frac{3x}{3} = \frac{30}{3}$

$x = 10$

Now, substitute the value of 10 into the expression for the angle measure that is $4x + 30$, to get $4(10) + 30 = 40 + 30 = 70°$. If you stopped at solving the equation and forgot to substitute into the angle measure expression, you would have chosen answer (A), a common mistake. If you had thought that the angle and the arc measures were supplementary, you would have set up the incorrect equation $4x + 30 + 7x = 180$, solved for x, and chosen answer (B). Choice (C) is the complement to the correct angle measure.

3. C

To convert from degree measure to radian measure, use the proportion
$\frac{\text{deg}}{radians} = \frac{360}{2\pi}$.

Let x represent the radian measure, and substitute in 270 for the degree measure.

The equation is: $\frac{270}{x} = \frac{360}{2\pi}$

Cross multiply to get: $540\pi = 360x$

Divide both sides by 360: $\frac{540\pi}{360} = \frac{360x}{360}$

Simplify the fractions to get: $\frac{3\pi}{2} = x$

Choice (A) can reflect a careless error in that you forgot to include the constant π in your answer. If your choice was (B), you may have simplified incorrectly. Choice (D) is the radian measure of the whole circle. Choice (E) is the value of the numerator of the radian measure.

4. E

When a diameter is perpendicular to a chord in a circle, the diameter bisects the chord. Therefore, the two segments XV and VZ are of equal length.

So, set up the equation: $5x + 10 = 3x + 130$

Subtract $3x$ from both sides: $5x + 10 - 3x = 3x + 130 - 3x$

Combine like terms: $2x + 10 = 130$

Subtract 10 from both sides: $2x + 10 - 10 = 130 - 10$

Combine like terms: $2x = 120$

Divide both sides by 2: $\frac{2x}{2} = \frac{120}{2}$

$x = 60$

Now, finish up the problem by finding the length of XZ. The measure of XZ is the sum of the two expressions: $5x + 10 + 3x + 130 = 8x + 140$. Substitute in 60 for x in the expression, to get $8(60) + 140 = 480 + 140 = 620$ units.

If your answer was choice (A), you may have set up the incorrect equation $5x + 10 + 3x + 130 = 180$, and then just solved for x. Choice (B) reflects a common error, in that this is the value of x, not the length of the chord. Choice (C) is just the sum of the numerical parts of the expressions that make up chord XZ. Choice (D) is the length of either XV or VZ.

5. A

When two chords intersect, the product of the segment parts of one of the chords is equal to the product of the segment parts of the other chord. In this figure, (EI)(IG) = (FI)(IH). Substitute in the given measures, and let EI be represented as x.

The equation is: $36x = (30)(60)$

Multiply on the right hand side: $36x = 1800$

Divide both sides by 36: $\frac{36x}{36} = \frac{1800}{36}$

$x = 50$

The length of EI is 50 units.

If you chose choice (B), you may have thought that EI would be twice the measure of IG. Choice (C) is the length of chord EG. Choice (D) is the length of chord FH. Choice (E) could be incorrectly arrived at simply by adding the three given measures.

6. C

When two chords intersect in a circle, the measure of the angle formed is one-half the sum of the intercepted arcs. In this problem, you are given the angle measure to be 55°, and one of the arc measures as 20°. Let x represent the missing arc measure, and:

Set up the equation: $55 = \frac{1}{2}(x + 20)$

Multiply both sides by 2: $55 \bullet 2 = 2 \bullet \frac{1}{2}(x + 20)$

Simplify to get: $110 = x + 20$

Subtract 20 from both sides: $110 - 20 = x + 20 - 20$

$90 = x$

So arc AD measures 90°.

Choice (A) is the measure of arc BC. If your choice was (B), you incorrectly thought that the arc measure would equal the angle measure. This is only true if the angle is a central angle. Choice (D) is twice the value of the angle measure. This relationship is only true for inscribed angles. If your choice was (E) you may have set up the equation incorrectly as the angle measures equal to one half of the difference between the arc measures, instead of the sum of the arc measures.

7. A

The measure of an angle formed by two tangent segments is one-half of the difference between the arcs that are intercepted. In the case of two tangents, the two arcs make up the entire circle with measure of 360°. Only one arc measure is given, that of arc JL equal to 130°. So arc JML is 360 – 130 = 230°. Now that you have the two arc measures, let x represent the angle measure, and $x = \frac{1}{2}(230 - 130)$, or $x = \frac{1}{2}(100) = 50°$.

If your choice was (B), you thought the angle would be one-half of the given arc measure. Choice (C) is the measure of arc JL. If you chose (D), you incorrectly thought the angle would be one-half of the sum, instead of the difference of the arc measures. Choice (E) is the arc measure of JML.

8. C

When two secants meet in the exterior of a circle, the product of the outer segment of one secant and the whole secant is equal to the product of the outer segment of the other secant and the whole secant. The whole length of secant QS is represented as SR + QR, or $x + 40$. Use these facts to set up an equation of (PQ)(NQ) = QR(SR + QR).

The equation is: $(50)(92) = 40(x+40)$

Multiply on the left hand side: $4600 = 40(x+40)$

Divide both sides by 40: $\frac{4600}{40} = \frac{40(x+40)}{40}$

Simplify to get: $115 = x + 40$

You do not need to solve the equation any further; the expression $x + 40$ represents the length of QS, equal to 115 inches.

If you chose answer choice (A), you made the incorrect assumption that the outer segment of QS would be 8 inches longer than the inner segment, as is the case for secant NQ. This is not a relationship that is true for secants. Choice (B) is the value of the variable x, or the length of segment QR. If you chose (D), you simply added the three measures given in the problem. Choice (E) could result from an incorrect solving process. On the right-hand side of the equation, you may have incorrectly thought that 40(40 + x) = 1,600 + x, instead of the correct use of the distributive property, which would give you 40(40 + x) = 1,600 + 40x.

9. D

The measure of the angle formed by two tangents to a circle, in this case ∠TVX, is one-half of the difference between the intercepted arcs. The figure gives the measure of the larger arc, arc TYX, but not the smaller arc, UW. But recall that a circle has degree measure of 360°, so the measure of UW is 360 – (20 + 20 + 240) = 360 – 280 = 80°. Now you can find the measure. The measure of $\angle TVX = \frac{1}{2}(240 - 80) = \frac{1}{2}(160) = 80°$.

Answer choice (A) is one-half of one of the outer arcs to the tangents. Choice (B) is the measure of one of these outer arcs. If your answer was (C), you may have been thinking of inscribed angles and incorrectly divided the arc measure UW by 2. Answer choice (E) may have been arrived at if you forgot to take one-half of the difference in the intercepted arcs, or if you calculated one-half of the sum of the arcs instead of one-half of the difference.

10. B

When a tangent and a secant meet in the exterior of a circle, the length of the tangent squared is equal to the product of the outer segment of the secant and the whole secant. Use this fact to solve for the missing tangent length. Let x represent the length of the tangent segment AB. The relationship is $(AB)^2 = (BC)(BD)$. The length of BD = BC + CD = 16 + 20 = 36.

Set up the equation: $x^2 = (16)(36)$

Multiply on the right hand side: $x^2 = 576$

Take the square root of both sides: $\sqrt{x^2} = \sqrt{576}$

$x = 24$

The length of tangent AB = 24 mm.

Choice (A) represents the length of segment CD. If your answer was choice (C), you found the length of secant BD. If you chose choice (D), you fell into a common trap; you divided both sides of the equation by 2 instead of taking the square root. Answer choice (E) is the value of the tangent length squared, not the tangent length. You may have forgotten that the theorem is used with the tangent value squared, not just the tangent length.

CHAPTER 5

Perimeter and Area of Polygons

WHAT ARE THE PERIMETER AND AREA OF A POLYGON?

This chapter will explore two basic and important concepts in geometry—perimeter and area. The *perimeter* is the distance around a polygon. This measure is used for many applications, such as fencing around a property or edging a garden. *Area* is the number of square units it takes to cover a polygon. When you want to paint your walls or cover a floor with carpeting, you need to calculate the area of the surface.

CONCEPTS TO HELP YOU

Your study of these measures will begin with an introduction to the basic concepts needed to tackle problems concerning perimeter and area. In addition to learning how to calculate these measures with basic polygons, this section will go on to explain how to find the area of an irregular shaped polygon and the shaded portions of polygons.

Perimeter of Various Types of Polygons

As stated earlier, the perimeter is the distance around a polygon. To find the perimeter, just add up all of the measures of the sides. Perimeter is therefore an addition concept. For example, to find the perimeter of the pentagon below, just add up the given measures.

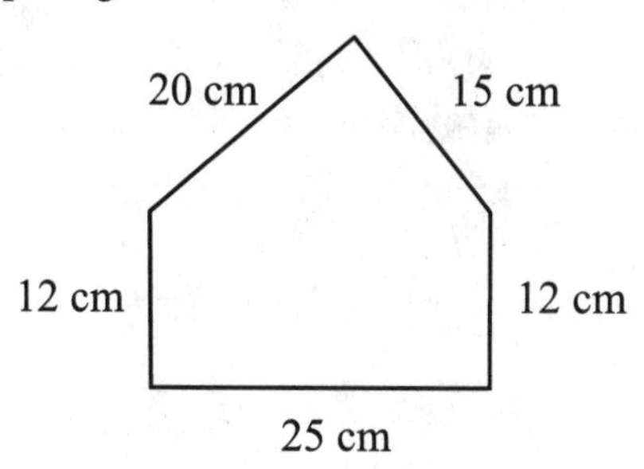

The perimeter is 20 + 15 + 12 + 25 + 12 = 84 cm. Note that perimeter is a linear measure so the units in this case is just centimeters. The distance around a circle is a special application of perimeter called the *circumference*. There are formulas that can be used to find the perimeter of the triangle and parallelogram and the circumference of a circle. These formulas will be given in the next section. When solving word problems that involve measurement, be sure you know when to use perimeter and when to use area. It is a common mistake to confuse the two concepts. Remember that perimeter is used when finding the distance *around* a figure.

Area of Various Types of Polygons

Area is the measure that is the number of square units it takes to cover a polygon. For example, the rectangle below has an area of 12 square units—there are 12 squares that can be counted in the rectangle. As was true for perimeter, there are formulas used to calculate area that will be illustrated in the next section.

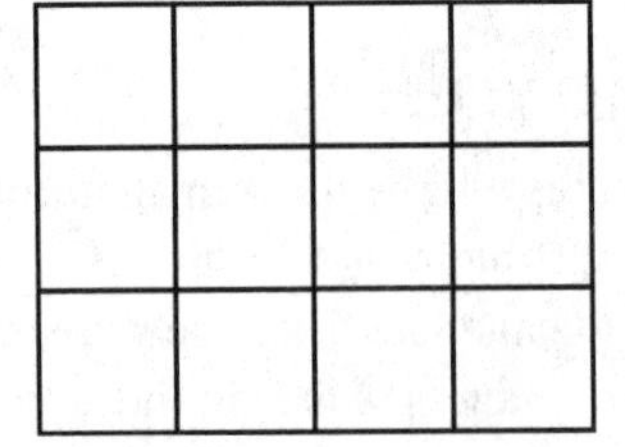

Most often, you need to use formulas to calculate area. Perimeter was an addition concept; area is a multiplication concept. The units used to measure area are square units. When solving word problems that involve measurement, be sure you do not confuse area with perimeter. Area is used when you need to *cover* a figure.

Irregular Shaped Polygons

Often, you are given figures that are not the standard polygons, such as triangles, parallelograms, trapezoids, or circles. These unusual figures are combinations of the basic shapes, and they are called *irregular figures*. Some examples are shown below:

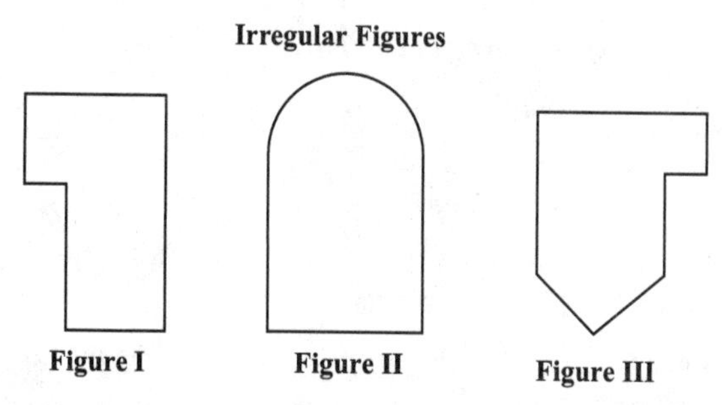

To find the perimeter of an irregularly shaped figure, just add up all of the side measures as you do for the basic polygons. If there are pieces of circles, such as shown above in Figure II, determine what fraction of the circle is present. In Figure II above, there is one-half of a circle. To find that part of the perimeter, you calculate the circumference, using the formula explained in the next section, and then divide it by 2. Then this is added to the three other side measurements.

To find the area of an irregularly shaped figure, break the figure up into the basic polygons, find the area of each part, and then add up the areas to get the total. The following irregular figures are shown as divided into basic shapes.

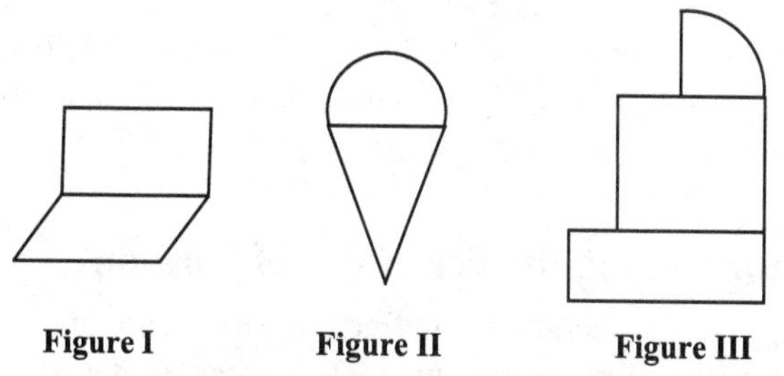

In the figures above, Figure I contains both a parallelogram and a rectangle. The total area will be the sum of the areas of each one. Figure II shows the figure broken up into a triangle and one-half of a circle. The total area will be the area of the triangle, plus the area of the circle divided by two. Figure III is a rectangle, a square, and one-fourth of a circle. Again, find the area of each piece and then add them together for the total area. To find the area of the piece of the circle, find the area of a whole circle and divide it by 4.

Finding the Area of Shaded Regions

Another common challenge involving area is finding the area of a shaded region, such as in the figures shown below.

Shaded Regions in Polygons

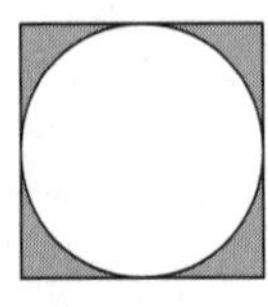

Figure I

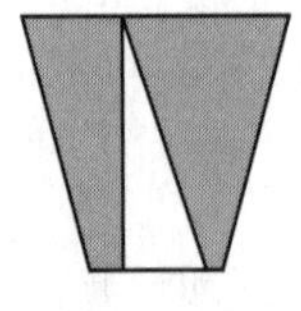

Figure II

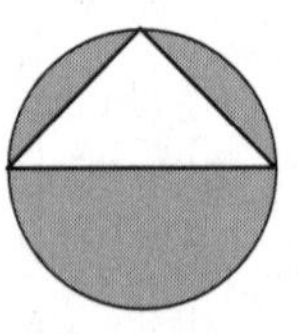

Figure III

The key to finding the area of the shaded regions is to first identify the basic shapes in the figure. There is an outer portion that is shaded and an inner portion. Figure I is a circle within a square. Figure II is a triangle within a trapezoid, and Figure III is a triangle within a circle. To find the area of the shaded region, calculate the area of the outer portion, calculate the area of the inner portion, and find the positive difference between these areas. $\text{Area}_{\text{outer piece}} - \text{Area}_{\text{inner piece}} = \text{Area}_{\text{shaded region}}$. For Figure I above, this is basically a square with the circle portion removed. Likewise, Figure II is a trapezoid with the triangular area removed, and Figure III is a circle with the triangular area removed.

STEPS YOU NEED TO REMEMBER

The previous section outlined the fundamental concepts involved with perimeter and area. This section will now introduce formulas and methods needed to tackle questions dealing with these measures.

Formulas for Finding the Perimeter of Polygons

You learned in the last section that to find the perimeter of a polygon, you can just add up all of the side measures. The perimeter of a circle is called the circumference, and since it is a curved surface, you must use a formula to calculate this measure. Because of the common use of the triangle in geometry, and the characteristics of the rectangle, square, and circle, there are formulas to calculate perimeter, as shown in the following illustration .

Perimeter of Four Basic Geometric Objects

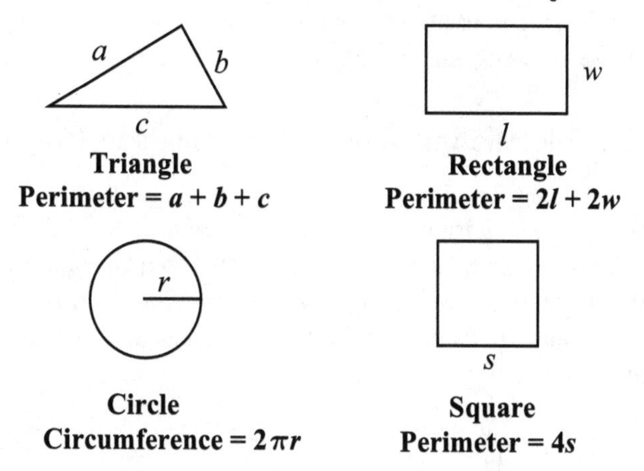

Use the formula shown on page 120 to find the perimeter of the rectangle below:

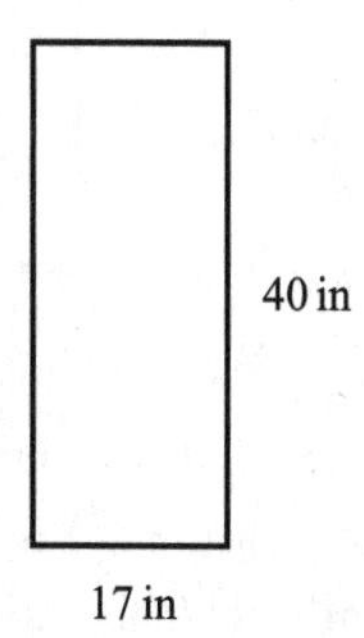

Use the formula and substitute in the given numbers for l and w. The perimeter is $2(40) + 2(17) = 80 + 34 = 114$ inches.

To find the circumference of the circle below, use the formula $C = 2\pi r$.

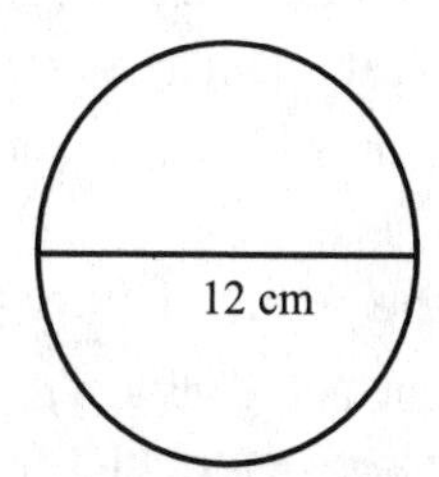

In this case, the radius is one half of the diameter of 12 cm, which is 6 cm. $C = 2 \cdot \pi \cdot 6 = 12$ cm. Often, answers dealing with circles have measures left in terms of the constant pi (π). If a numerical answer is required that is not in terms of π, be sure to use the π key on your calculator, and then round the answer to the correct decimal place.

How to Handle Missing Side Measurements of Irregular Figures

When presented with an irregular figure and asked to find the perimeter or area, sometimes there are missing side measures and it appears that there is not enough information. Look closely at the figure, however. You must often use logical reasoning to determine missing side measures. For example, examine the irregular figures below.

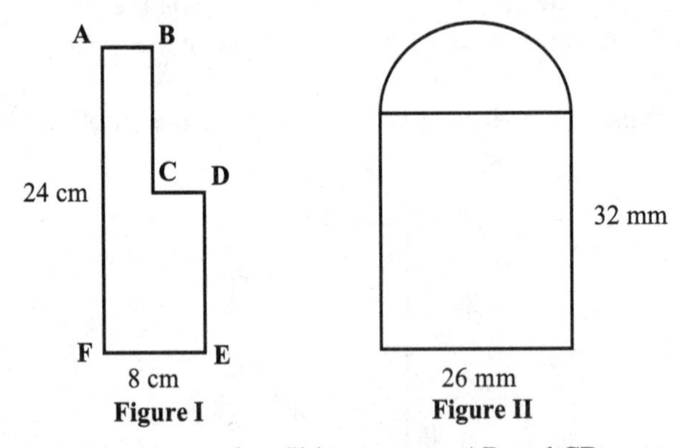

Figure I **Figure II**

Figure I contains two rectangles. Side measures AB and CD are not given, but they will sum to the same measure as FE. Likewise, the unknown side measures BC and DE will sum to the same measure as AF. So the perimeter is $8 + 24 + 8 + 24 = 64$ cm. In Figure II, the radius of the half circle is unknown. But the diameter is the same measure as the width of the rectangle that is 26 mm; the radius is one half of this, or 13 mm. Also note that the missing length is equal to the given length of 32 because the bottom portion of the figure is a rectangle. So the perimeter of this figure is $32 + 26 + 32 + \frac{1}{2}(2 \bullet \pi \bullet 13) = 90 + 13\pi$ mm in length.

Formulas for Finding the Area of Polygons

The characteristics of the basic polygons and the circle were described in previous chapters. There are known formulas for determining the area of these common figures, as illustrated on the next page.

Area Formula for Common Geometric Figures

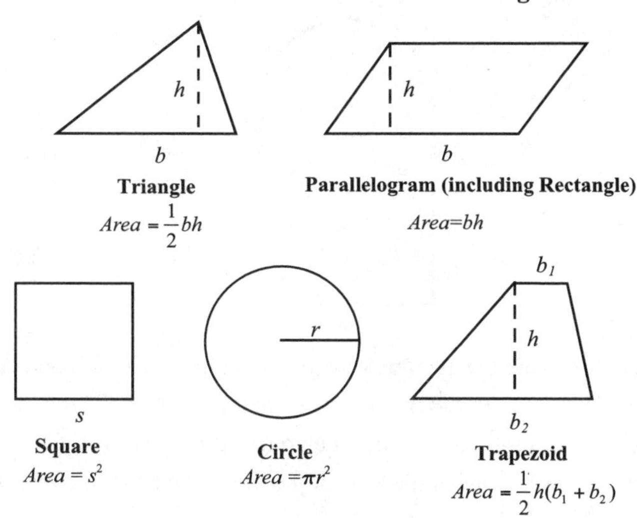

For example, to find the area of the trapezoid below, substitute the given measures into the area formula.
The area is $\frac{1}{2} \bullet 76 \bullet (80 + 62) = 38 \bullet 142 = 5396$ mm^2.

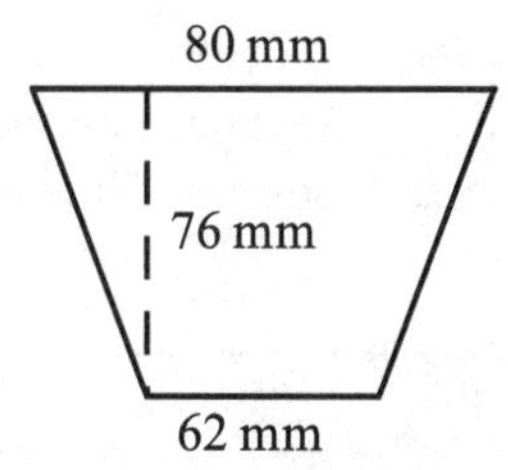

To find the area of a circle with a radius of 12 cm, use the formula $Area = \pi r^2$. The area of the circle is $\pi \bullet 12^2 = 144\pi$ cm^2.

Finding the Area of Irregular Figures

To find the area of an irregular figure, first break up the figure into recognizable geometric shapes. Once this is complete, find the area of each piece. The area of the irregular figure will be the sum of these individual shapes. As explained in the last section, sometimes irregular figures will have parts of a circle, such as a half, a quarter, or three quarters of a circle. In those cases, compute the area of the whole circle and then multiply by either one-half, one-fourth, or three-fourths depending on the figure.

For example, find the area of the irregular figure below:

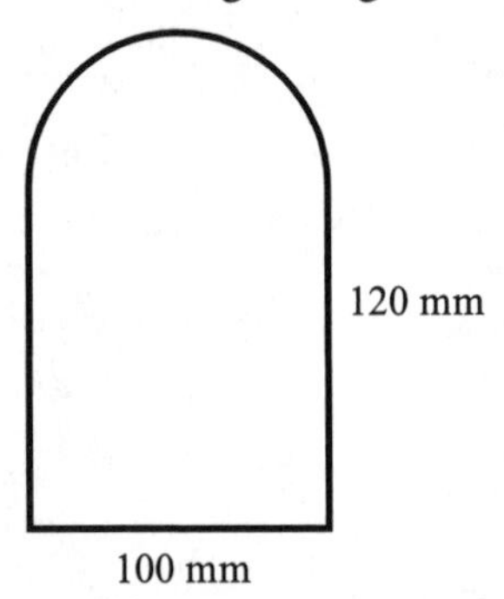

The irregular figure is a rectangle with one half of a circle. The diameter of the circle is equal to the width of the rectangle; that is 100 mm. So the radius of the circle is 50 mm. The area of this figure is the area of the rectangle, plus one-half of the area of the circle. Use the formulas to get $Area = bh + \frac{1}{2}\pi r^2$. Substituting in the given measures you get $Area = (100 \bullet 120) + \frac{1}{2} \bullet \pi \bullet 50^2 = 12,000 + \frac{1}{2} \bullet \pi \bullet 2,500 = 12,000 + 1,250\pi$ mm^2.

Finding the Area of Shaded Regions

To find the area of a shaded region, identify the outer and inner geometric shapes. Then, find the positive difference between the shapes. For example, the figure below is an outer parallelogram with an inner triangle. The shaded area surrounds the triangle.

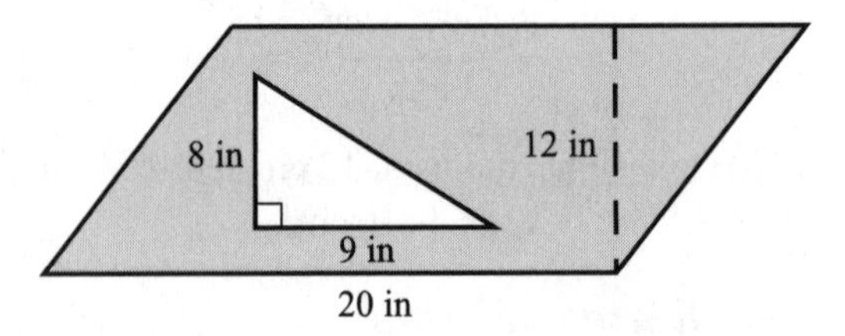

The area of the shaded region is $\text{Area}_{\text{parallelogram}} - \text{Area}_{\text{triangle}}$. Substituting in the formulas gives the area as $b_1h_1 - \frac{1}{2}b_2h_2$, where b_1 and h_1 are the base and height of the parallelogram and b_2, h_2 are the base and height of the triangle.

Now, using the given dimensions, the area of the shaded region is $20 \bullet 12 - \frac{1}{2} \bullet 9 \bullet 8 = 240 - 36 = 204$ in^2.

Finding a Side Measure if Perimeter or Area Is Given

If you are given the perimeter, or area, and missing side measurements, you will use the formula and algebra to solve for the missing side. For example, if the perimeter of the rectangle below is 458 cm, what are the dimensions of the rectangle?

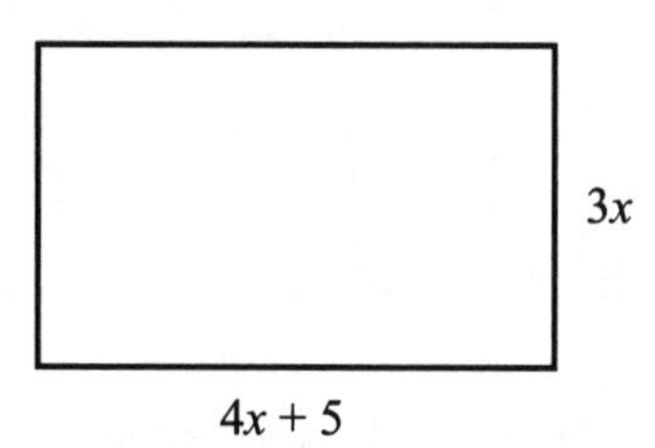

Use the formula for the perimeter of a rectangle to set up an equation:	$2(4x + 5) + 2(3x) = 458$
Use the distributive property on the left-hand side:	$8x + 10 + 6x = 458$
Combine like terms:	$14x + 10 = 458$
Subtract 10 from both sides:	$14x + 10 - 10 = 458 - 10$
Simplify and divide both sides by 14:	$\frac{14x}{14} = \frac{448}{14}$
	$x = 32$

Now, substitute in the value of 32 for x in the expressions to find the dimensions. The width, $3x$, is $3(32) = 96$ cm, and the length, $4x + 5$, is $4(32) + 5 = 128 + 5 = 133$ cm.

Similarly, if the area of a circle is 289π mm^2, what is the diameter of the circle?

Using the formula, the equation is:	$\pi r2 = 289\pi$
Divide both sides by π:	$\frac{\pi r^2}{\pi} = \frac{289\pi}{\pi}$
The equation is now:	$r^2 = 289$
Take the square root of both sides:	$\sqrt{r^2} = \sqrt{289}$
	$r = 17$

If the radius is 17 mm, then the diameter is $2 \bullet 17 = 34$ mm.

STEP-BY-STEP ILLUSTRAION OF THE FIVE MOST COMMON QUESTION TYPES

Use this section to expand your experience with perimeter and area type questions. These problems represent common problem situations that you may face in real-life situations, or on various types of standardized tests.

Question 1: The Effects of Doubling the Dimensions of a Polygon

If you double both dimensions of a square, what happens to the perimeter and the area?

(A) The perimeter and area do not change

(B) The perimeter and the area are 2 units bigger

(C) The perimeter and the area will double

(D) The perimeter and the area will be four times larger

(E) The perimeter will double and the area will be four times larger

In order to understand this problem, make a picture with a square with sides of length x, and then a picture of the square with the sides doubled:

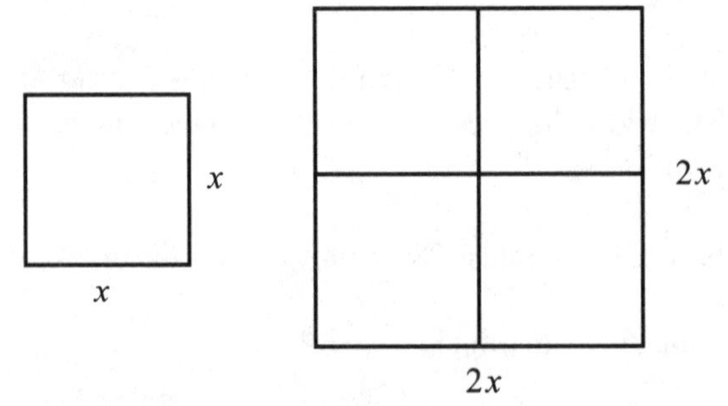

From the picture, you can see that the area will be four times larger; there are four of the smaller squares contained in the larger square. Or, using the formula, the area of the small square is x^2 and the area of the large square is $(2x)^2 = 4x^2$, four times bigger. To find what happens to the perimeter, calculate the perimeter for each. The perimeter of the small square is $4x$. The perimeter of the new square with dimensions doubled is $4(2x) = 8x$, which is twice as big; therefore, the perimeter doubled. **Choice (E) is the correct answer.**

To say that a dimension is 2 units bigger as in choice (B), would be to make the dimensions x plus 2, not doubling, which is x times 2. If you did not draw a picture, a common error to be made would be to assume that both the perimeter and the area doubled, choice (C).

Question 2: Finding Dimensions of a Polygon Given the Ratio and the Area

The base and height of a triangle are in the ratio of 4:3. If the area of the triangle is 150 mm^2, what is the measure of the height?

(A) 5 mm

(B) 10.6 mm

(C) 14.14 mm

(D) 15 mm

(E) 20 mm

If the base and height are given as the ratio of 4:3, multiply each part of the ratio by the variable x, to get an equivalent ratio of $4x:3x$. This means that the base is represented as $4x$ and the height as $3x$. Use the formula for the area of a triangle, $Area = \frac{1}{2}(base)(height)$, and solve for the variable x.

The equation is: $150 = \frac{1}{2} \bullet 4x \bullet 3x$

Multiply on the right hand side: $150 = 6x^2$

Divide both sides by 6: $\frac{150}{6} = \frac{6x^2}{6}$

The equation is now: $25 = x^2$

Take the square root of each side: $\sqrt{25} = \sqrt{x^2}$

$5 = x$

Substitute the value of x into the expression for the height, $3x$. The height is $3(5) = 15$ mm, so **choice (D) is the correct answer.**

If you chose (A), you fell into the common error of giving the value of the variable x, instead of the height of the triangle. If you used an incorrect formula for a triangle, such as Area = (base)(height), instead of the correct formula, but did all other steps correctly, you would arrive at answer (B) for the height. Choice (C) is the length of the base of the triangle using that same incorrect formula as described for choice (B). Choice (E) is the correct length of the base of the triangle, not the height; another common mistake made when solving these types of problems.

Question 3: Finding a Missing Dimension Given the Area of a Polygon

What is the length of base AD in the trapezoid below, if the area is 132 in^2?

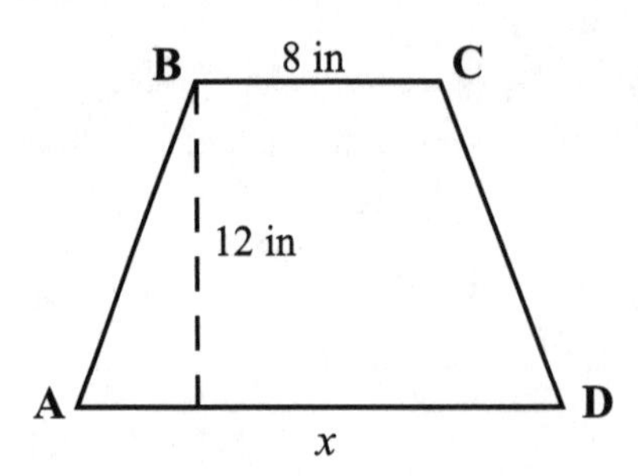

(A) 2.75 in

(B) 3 in

(C) 14 in

(D) 20.67 in

(E) 22 in

Use the formula for the area of a trapezoid, that is $Area = \frac{1}{2}(height)(base_1 + base_2)$, and solve the equation for the missing base, x.

The equation is:	$132 = \frac{1}{2} \bullet 12(x+8)$
Multiply on the right hand side:	$132 = 6(x+8)$
Use the distributive property on the right hand side:	$132 = 6x + 48$
Subtract 48 from both sides:	$132 - 48 = 6x + 48 - 48$
Simplify to get	$84 = 6x$
Divide both sides by 6:	$\frac{84}{6} = \frac{6x}{6}$
	$14 = x$

The missing base length is 14 in., so choice **(C) is the correct answer.**

If you chose (A), you may have arrived at the incorrect equation $132 = 48x$. If your answer was choice (B), you forgot the factor in the formula. If you chose (D), you may have forgotten the distributive property and had the incorrect equation of $132 = 6x + 8$, instead of $132 = 6(x + 8)$. Choice (E) would be arrived at by using the incorrect formula as Area = (base)(height) with the base of x and the height of 12.

Question 4: Finding the Area of an Irregularly Shaped Polygon

What is the area of the figure below?

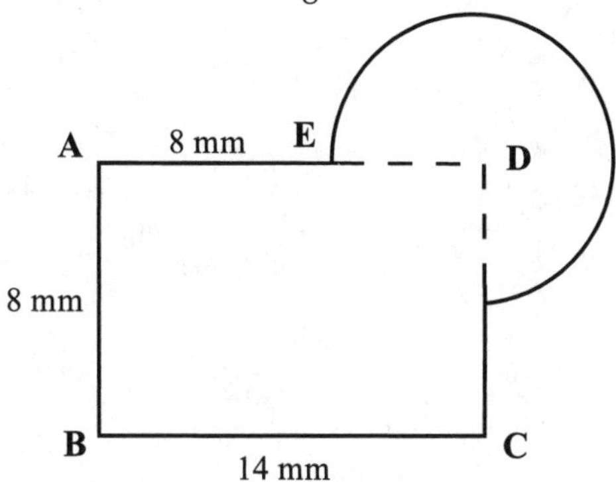

(A) 112 mm^2

(B) 139 mm^2

(C) 148 mm^2

(D) $(27\pi + 112)$ mm^2

(E) $(36\pi + 112)$ mm^2

The irregular figure shown is a rectangle and three quarters of a circle. Use the area formulas and find the area of each piece and add them together. The area of a rectangle is *base* times *height*, or 8(14) = 112 mm^2. The area of the circle piece is $\frac{3}{4}\pi r^2$. The radius of the circle is BC – AE, or 14 – 8 = 6 mm. So the area of the circle is $\frac{3}{4} \bullet \pi \bullet 6^2 = 27\pi$ mm^2. The total area of the figure is $(27\pi + 112)$ mm^2, which makes choice **(D) the correct answer.**

Choice (A) is the area of the rectangular portion only. If you forgot to include the constant π, you may have selected choice (B), or 27 + 112. Choice (E) is the area of the rectangle, plus a whole circle, not the three quarters. If you added the area of the rectangle to the area of the whole circle, and forgot the constant π, you would have chosen choice (C), or 36 + 112.

Question 5: Finding the Area of a Shaded Region

Find the area of the shaded region:

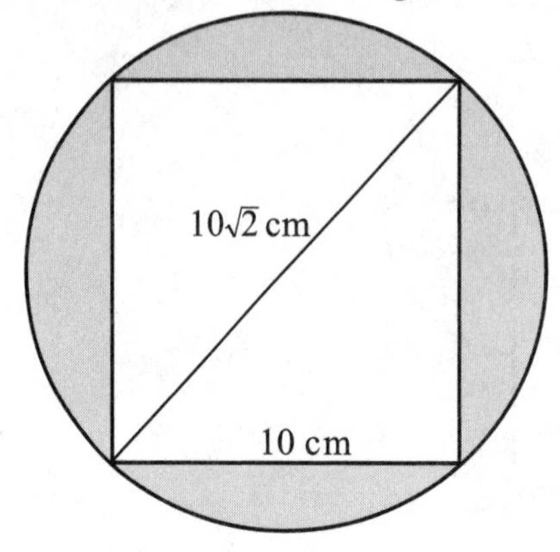

(A) $(50\pi - 100)$ cm^2

(B) $(100 - 50\pi)$ cm^2

(C) $(200\pi - 100)$ cm^2

(D) $(100 - 200\pi)$ cm^2

(E) $(400\pi - 100)$ mm^2

To find the area of the shaded region, first determine that this is a square within a circle. The shaded area will be the area of the circle, minus the area of the square. The formula for the area of a circle is πr^2. It seems as though the radius is not given in the problem, but notice that the diagonal of the square, of length $10\sqrt{2}$ cm, is also a diameter of the circle. The radius will be half of this diameter length, or $5\sqrt{2}$ cm long. The sides of the square measure 10 cm as shown in the figure. The area of the shaded region is therefore $\pi(5\sqrt{2})^2 - 10^2 = (50\pi - 100)$ cm^2, so choice (A) is the correct answer.

Choice (B) is the area of the square, minus the area of the circle, and would result in a negative value. If you chose (C), you mistakenly considered the diameter length to be the radius. Choice (D) is the area of the square, minus the area of a circle of diameter $10\sqrt{2}$ in length. Choice (E) would represent the area of a circle with radius of 20 cm, minus the area of the square.

CHAPTER QUIZ

1. If the perimeter of the following hexagon is 98 inches, what is the value of x?

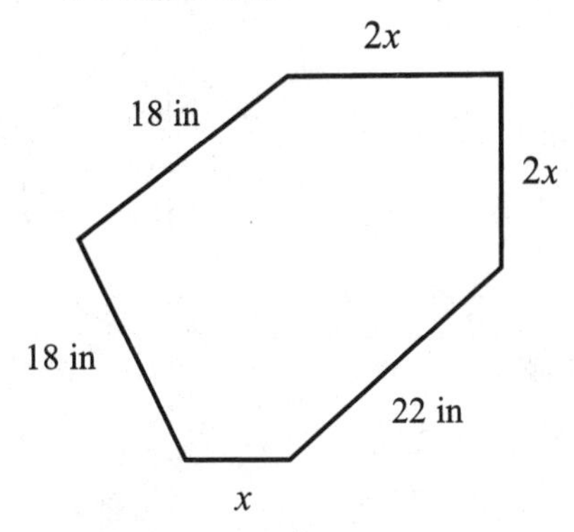

(A) 8 in

(B) 16 in

(C) 19.6 in

(D) 31.2 in

(E) 40 in

2. If you triple the radius of a circle, what happens to the circumference and the area?

(A) There will be no change.

(B) The area and the circumference will both triple.

(C) The circumference will triple and the area will be six times larger.

(D) The circumference and the area will be six times larger.

(E) The circumference will triple and the area will be nine times larger.

3. What is the area of trapezoid WXYZ below?

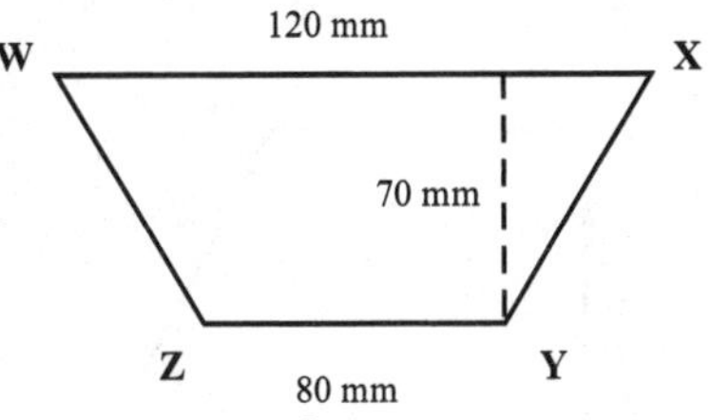

(A) 270 mm^2

(B) 2,800 mm^2

(C) 5,600 mm^2

(D) 7,000 mm^2

(E) 14,000 mm^2

4. Given that the length and width of a rectangle are in the ratio of 7:4, what is the width of the rectangle if the area is 448 cm^2?

(A) 4 cm

(B) 16 cm

(C) 28 cm

(D) 32 cm

(E) 162.88 cm

5. Given that the base of a triangle is 62 m, if the area is 1,240 m^2, what is the height?

(A) 10 m

(B) 20 m

(C) 40 m

(D) 80 m

(E) 1,178 m

6. If the area of trapezoid LMNO below is 870 cm², what is the length of side LM?

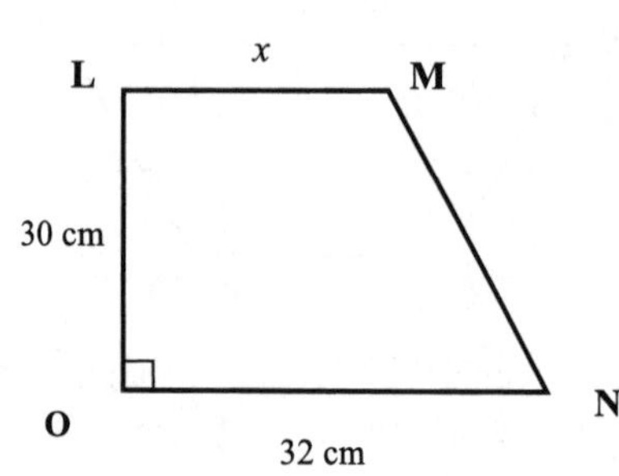

(A) 26 cm
(B) 29 cm
(C) 30 cm
(D) 55.87 cm
(E) 58 cm

7. What is the perimeter of the figure below?

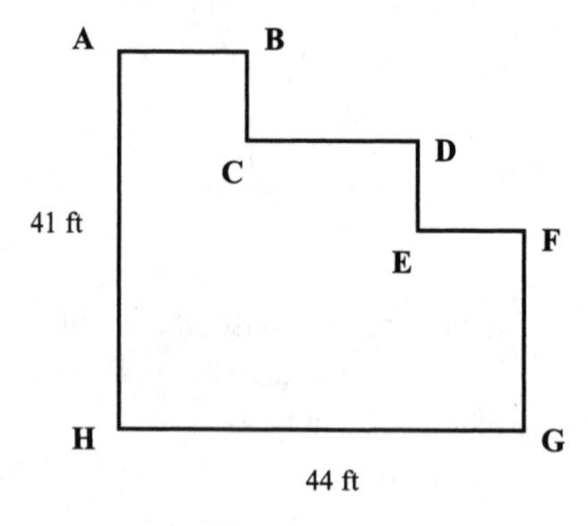

(A) 42.5 ft
(B) 85 ft
(C) 126 ft
(D) 170 ft
(E) 1,804 ft

8. If the triangular portion of this irregular figure is an isosceles triangle, what is the area of the entire figure?

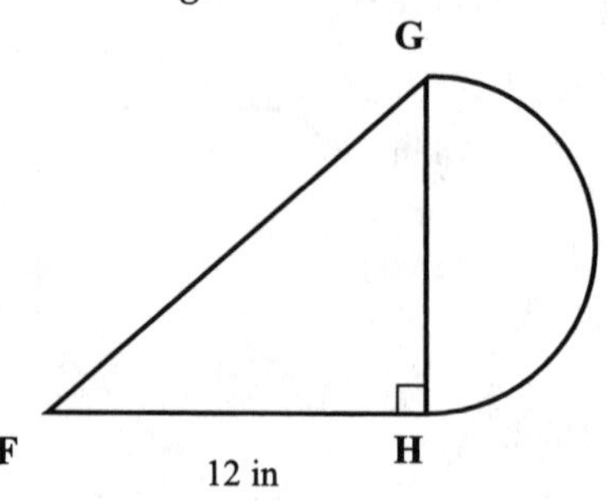

(A) 72 in²
(B) $(72 + 18\pi)$ in²
(C) 144 in²
(D) $(72 + 36\pi)$ in²
(E) $(144 + 18\pi)$ in²

9. What is the area of the shaded region in the figure below?

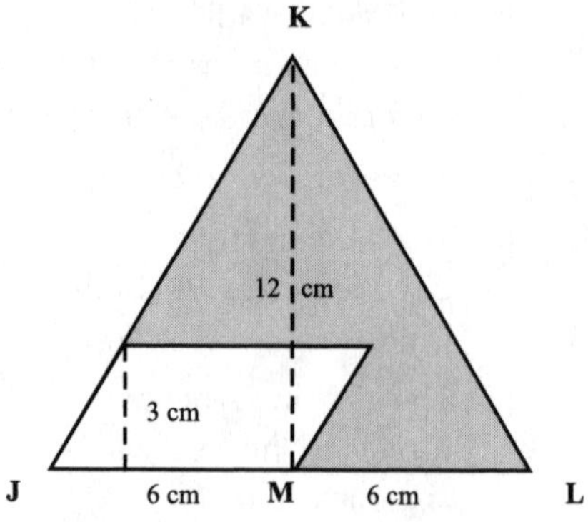

(A) 18 cm²
(B) 54 cm²
(C) 63 cm²
(D) 126 cm²
(E) 144 cm²

10. The following figure is a circle inscribed in a square. What is the area of the shaded region?

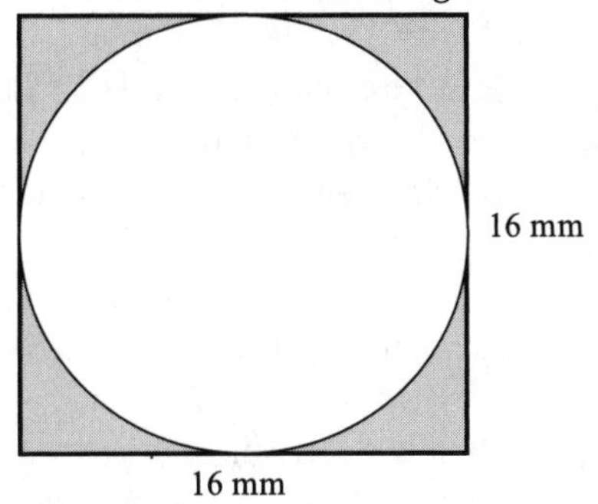

(A) 256 mm^2
(B) $(32 - 16\pi)$ mm^2
(C) $(256 - 64\pi)$ mm^2
(D) $(32 + 16\pi)$ mm^2
(E) $(256 + 64\pi)$ mm^2

ANSWER EXPLANATIONS

1. A

The perimeter is the distance around a polygon. You are given the perimeter, and are asked to find the value of the variable x.

Set up an equation: $2x + 2x + x + 18 + 18 + 22 = 98$

Combine like terms on the left hand side: $5x + 58 = 98$

Subtract 58 from both sides: $5x + 58 - 58 = 98 - 58$

Simplify and divide both sides by 5: $\frac{5x}{5} = \frac{40}{5}$

$x = 8$

If your answer was choice (B), you found the value of the expression $2x$, or $2(8) = 16$. If you ignored the numerical sides and set up the incorrect equation as $5x = 98$, your answer would have been choice (C). If you had added 58 to both sides of the equation in the second step, instead of subtracting 58, your answer choice would be (D). Choice (E) is the value of $5x$, not the value of x.

2. E

To figure out what happens to a circle when you triple the radius, find the area and circumference of a circle with a radius of x, and then do the same for a circle with a radius of $3x$. A circle with a radius of x has a circumference of $2\pi x$ units. When the radius is $3x$, the circumference is $2 \bullet \pi \bullet 3x = 6\pi x$. Therefore, the circumference tripled. A circle of radius x has an area of πx^2 square units. When the radius is $3x$, the area is $\pi(3x)2 = 9\pi x2$. The area is thus nine times larger.

If you forgot to square the "3" coefficient in the area formula, your answer choice would be (B). If you chose (C), you may have made the common mistake of multiplying 3 times 2 to get 6, instead of squaring the 3 to get 9. If you chose (D), you may have thought that since the circumference formula was $2\pi r$, that the circumference would be six times as large.

3. D

Use the formula for the area of a trapezoid, which is

$$Area = \frac{1}{2} \bullet height(base_1 + base_2).$$

Substitute in the correct values to get

$$Area = \frac{1}{2} \bullet 70(120 + 80) = 35 \bullet w200 = 7,0\text{mm}^2.$$

If your answer was choice (A), you may have just added the three measures given. If you chose (B), you may have mistakenly multiplied $\frac{1}{2} \bullet 70 \bullet 80$ and ignored the second base of 120. Choice (C) is the area of a parallelogram with a base of 80 and a height of 70, not the area of the given trapezoid. Choice (E) is twice the area of the trapezoid; you forgot to multiply by one-half.

4. B

The length and width are given as the ratio of 7:4. You can multiply each part of the ratio by the variable x, to get an equivalent ratio of $7x$:$4x$. This means that the length is represented as $7x$ and the width as $4x$. Use the formula for the area of a rectangle, *Area = (base)(height)*, where the base is the length and the height is the width. Substitute in the value of the area, and the expressions for the length and width.

The equation is: $(7x)(4x) = 448$

Multiply on the left hand side: $28x^2 = 448$

Divide both sides by 28: $\frac{28x^2}{28} = \frac{448}{28}$

Simplify and take the square root of both sides: $\sqrt{x^2} = \sqrt{16}$

$x = 4$

The value of x is 4, so the width, $4x$, is $4(4) = 16$ cm.

Don't fall into the common trap of choice (A), which is the value of the variable, not the measure of the width. A second common mistake is choice (C), which is the measure of the length, not the width. If you chose (D) or (E), you divided 16 by 2 instead of taking the square root of 16, resulting in the incorrect value of $x = 8$. Choice (D) would be the incorrect value of the width with this error, and choice (E) the incorrect value of the length.

5. C

In this problem, you are given the area and the measure of the base of a triangle. To find the height, use the formula for the area of a triangle: $Area = \frac{1}{2} \bullet base \bullet height$. Let x represent the height, and substitute into the formula.

The equation is: $\frac{1}{2} \bullet 62 \bullet x = 1,240$

Multiply on the left hand side: $31x = 1,240$

Divide both sides by 31: $\frac{31x}{31} = \frac{1,240}{31}$

$x = 40$

The height is 40 m. If your answer was choice (A), you divided the area by the base, and then took one half of that answer. If you chose (B) you simply divided the area by the base, as if the geometric figure was a parallelogram instead of a triangle. Choice (D) is twice the height. Choice (E) would be arrived at if you simply subtracted the base from the area of the triangle.

6. A

To find the length of LM, use the formula for the area of a trapezoid. Let x represent the length LM. The trapezoid has a right angle at $\angle$LON, so side LO is the height of the figure. The formula for the area of a trapezoid is $Area = \frac{1}{2} \bullet height(base_1 + base_2)$.

Set up the equation: $\frac{1}{2} \bullet 30(x+32) = 870$

Multiply on the left hand side: $15(x + 32) = 870$

Divide both sides by 15: $\frac{15(x+32)}{15} = \frac{870}{15}$

Simplify: $x + 32 = 58$

Subtract 32 from both sides: $x + 32 - 32 = 58 - 32$

$x = 26$

The length of LM is 26 cm. If you chose (B), you thought you were working with a parallelogram with height of 30 and a base of LM. Therefore, you divided the area by 30 to get the incorrect value. Answer choice (C) is the length of LO, which is not congruent to LM. If your choice was (D), you came up with the incorrect equation of $870 = 15x + 32$; you used the distributive property incorrectly, and thought that $15(x + 32) = 15x + 32$; a correct use of the property would produce the equation $15x + 480 = 870$. If you forgot to add in the base of 32 to the original equation when using the area formula, your answer choice would be (E).

7. D

At first glance, it appears that there is not enough information given. But because this irregular figure is made up of rectangles, look closely and notice that the length of AH = 41, is equal to BC + DE + FG; Likewise, HG = 44, is equal to AB + CD + EF. So the perimeter is simply 41 + 44 + 41 + 44 = 85 + 85 = 170 ft.

If your choice was (A) you divided the sum of 41 and 44 by 2, instead of the correct calculation of multiplying by 2. If you chose answer choice (B), you just added the two numbers given in the figure. If your answer was (C), you added 41 + 41 + 44 and forgot to add in the other value of 44. Choice (E) would be the area of a large rectangle with base of 44 and height of 41.

8. B

You are told that the triangular portion of this irregular figure is isosceles, so you can conclude that the length of GH is congruent to FH, and therefore both have the measure of 12 inches. There is also a half circle; the diameter of the circle is GH, of measure 12, so the radius has a measure of 6 inches. Looking at the figure, angle $\angle$GHF is a right angle, so therefore the base of the triangular portion is FH and the height is GH. The total area of the figure will be the sum of the triangle, plus one half of the area of a circle. Use the

formulas for the area of a triangle and a circle to get the formula:
$Area = (\frac{1}{2} \bullet base \bullet height) + \frac{1}{2}(\pi r^2)$. Substitute in the values to get
$Area = (\frac{1}{2} \bullet 12 \bullet 12) + (\frac{1}{2} \bullet \pi \bullet 6^2) = 72 + 18\pi$ in^2.

Choice (A) is the area of the triangle only. If you chose answer choice (C) you just multiplied the base times the height of the triangle, did not take one-half, and ignored the half circle. Choice (D) results from a common error; you forgot to take one half of the circle, and gave the area of the triangle plus a whole circle. Answer choice (E) would result from an incorrect formula for the triangle (forgetting the one-half in the formula) and adding this to a whole circle.

9. B

The shaded region in this figure is the area of a triangle, minus the area of a parallelogram. The base of the triangle is 6 + 6 = 12 cm, and the height is also 12 cm. The parallelogram has a base of 6 cm and a height of 3 cm.Use the formulas for the area of a triangle, that is $Area = \frac{1}{2} \bullet base \bullet height$, and the area of a parallelogram, that is *Area* = *base* • *height*. Substitute in the values and find the difference between the areas. The area of the shaded region is $(\frac{1}{2} \bullet 12 \bullet 12) - (6 \bullet 3) = 72 - 18 = 54$ cm^2.

Choice (A) is the area of the parallelogram. If your answer was choice (C), you found the difference between the area of the triangle and one half the area of the parallelogram. Answer choice (D) would be arrived at if you forgot to take one-half of the product of the base and height of the triangle, but remembered to subtract out the area of the parallelogram. Answer choice (E) is the area of a large parallelogram with a base of 12 and a height of 12.

10. C

The area of the shaded region is the area of the square, minus the area of the circle. In order to solve this problem, you need to recognize that the diameter of the circle is equal to the side of the square, of length 16 mm. Therefore, the radius of the circle is 8 mm. The area of a square is *Area* = s^2, and the area of a circle is *Area* = πr^2. So the area of the shaded region is *Area* = 162 – π • 82 = (256 – 64π) mm^2.

Answer choice (A) is the area of the square. You may have chosen this answer if you did not recognize that the circle has a diameter equal to the side of the square, and therefore you chose to ignore the circle. If you thought that squaring meant to multiply by 2 instead of multiplying a value by itself, you would have chosen answer (B). Likewise with answer choice (D); it would be the same mistake, and in addition, you added the values instead of subtracting. If you chose answer (E), you fell into a common trap; you used the formulas correctly, but mistakenly found the sum of the areas instead of the difference.

CHAPTER 6

Surface Area

WHAT IS SURFACE AREA?

The *surface area* of a three-dimensional figure is the area of the outside covering of the object. In some cases, the surface area is relatively easy to find because each of the faces that comprise the outer covering are squares and rectangles. In other solids, such as cylinders, this outside covering is made up of circular objects, so finding the surface area is slightly more of a challenge. Either way, the formulas for each of the figures boil down to finding the area of each side, or face, and adding these areas together.

CONCEPTS TO HELP YOU

Begin your study with some basic terms of surface area. This section will then walk through the construction of some of the most common three-dimensional solids and the formula for finding the surface area of each.

Basic Terms of Surface Area

The concept of surface area cannot be discussed until a few terms are understood. These terms are *face, edge,* and *vertex.*

A *face* of a three-dimensional figure is the shape or shapes that make up the surface of the figure. A face that forms the top and/or bottom of a solid is called a *base*. The faces that are polygons and are not bases are called *lateral faces.*

An *edge* is the line segment that is formed when two faces meet.

The *vertex* is the place where two or more edges meet. You can think of the vertex as a "corner" of the three dimensional figure.

The location of each of these in a prism is shown in the figure below.

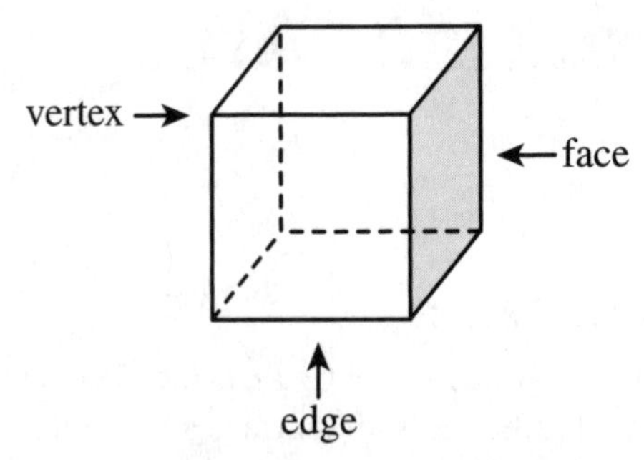

These three terms are used to define various three dimensional objects, as well as help calculate the surface area of each. Keep in mind that surface area is a two-dimensional calculation, so always use square units as a label. Let's start our study by describing each figure and how it is constructed.

SURFACE AREA AS A NET

To begin looking at the surface area of specific objects, start with the idea that each figure has an outer covering, or *net,* that can be visualized when calculating the surface area. The process of unfolding this outer covering can be very useful for understanding and remembering the formulas for finding the surface area of various figures.

Prisms

One of the most basic of all three-dimensional figures is the prism. A *prism* is a three-dimensional figure with congruent bases and lateral faces. The bases of any prism are parallel to each other. Each prism is classified by the type of bases it has.

A *rectangular prism* is a prism with rectangles as the congruent bases.

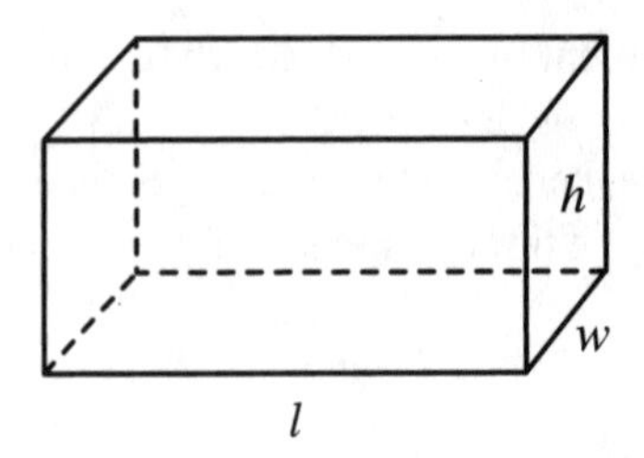

A *triangular prism* is a prism with triangles as the congruent bases.

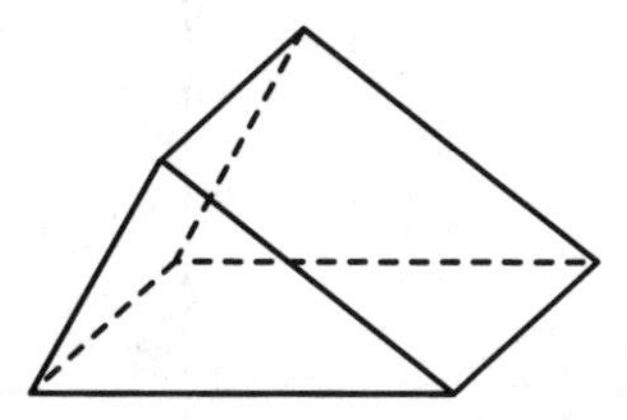

There are also other types of prisms and, as stated above, each is classified according to the shape of its congruent bases. For example, a prism with pentagons as bases is called a *pentagonal prism* and a prism with hexagons as bases is called a *hexagonal prism.*

The most common prism is the rectangular prism, as shown in the figure at the bottom of the previous page. There are six faces to a rectangular prism so the net of a rectangular prism consists of six rectangles. This net can be visualized like the figure below.

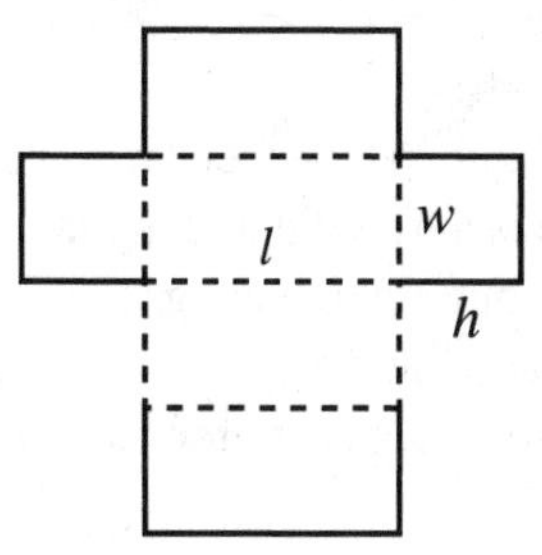

In any rectangular prism, the top and bottom bases are congruent, the right and left faces are congruent, and the front and back faces are congruent. Therefore this net consists of two faces where the area is equal to *length* • *width,* two faces where the area is equal to *length* • *height,* and two faces where the area is equal to *width* • *height.* The surface area formula can be summarized as $SA = 2lw + 2lh + 2wh$, where l is the length of the figure, w is the width, and h is the height.

Cubes

A cube is a special type of rectangular prism. A *cube* has six congruent faces, where each face is a square whose side is an edge of the cube. The figure below shows one way that the net for the cube can be drawn.

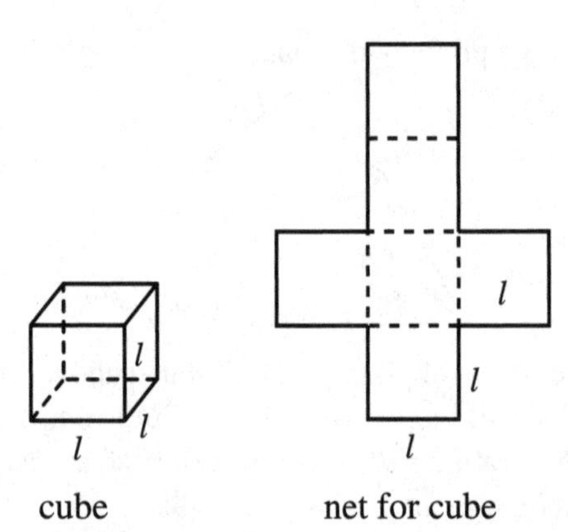

Since each face is a square, the area of a face can be found by multiplying *Area* = *edge* • *edge* or $A = e \bullet e = e^2$. Since there are six congruent squares that make up the surface area, the formula for the surface area of a cube is equal to $SA = 6e^2$.

Pyramids

Pyramids are similar to prisms in that they are defined by their base. However, each lateral face is a triangle and the lateral faces merge at a single point. So, instead of having two congruent bases, pyramids have one base, triangles for lateral faces, and a vertex point that commonly appears at the top of the figure. A square-base pyramid and its net are shown in the figure below.

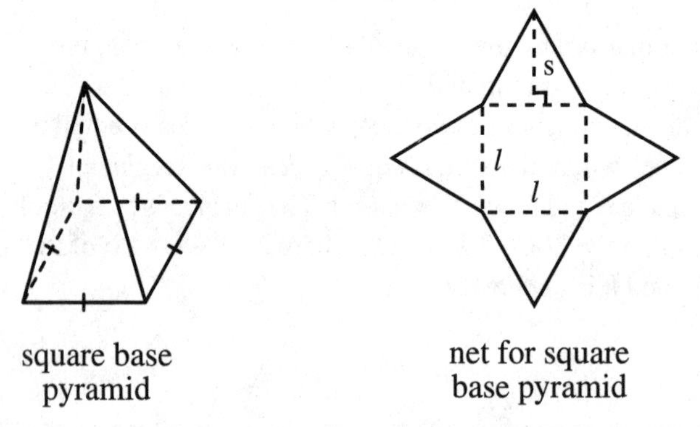

Be careful when finding the height of the lateral faces of a pyramid. For this figure, you want to use the slant height for the figure, not the height of the pyramid.

SLANT HEIGHT OF A PYRAMID

The *slant height* is the altitude of the triangular faces of the pyramid. It differs from the height of the pyramid itself because the slant height is not drawn from the vertex to a point in the center of the base of the pyramid. An example of each type of height is shown in the figure below.

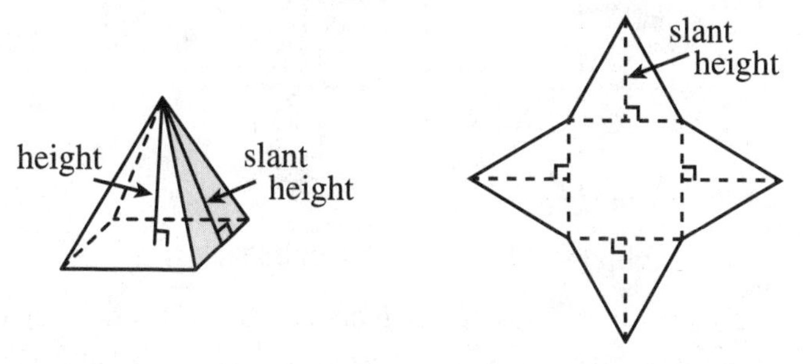

In a square-base pyramid, each of the sides of the square base has an edge of length e. The area of this square base is then $e \bullet e = e^2$.

Each lateral face is a triangle. Recall from Chapter 5 that the area of a triangle is equal to $A = \frac{1}{2}bh$, where b is the length of the base of the triangle and h is the height. The four triangular faces of a square-base pyramid each have a base that is equal to the edge of the square base and a height that is equal to the slant height of the pyramid. Thus, the area of one of these triangular faces is equal to $A = \frac{1}{2}es$, where e is an edge of the square base and s is the slant height. Since there are four congruent triangular faces, the surface area of all four is equal to $4(\frac{1}{2}es) = 2es$.

With these points in mind, the formula for the surface area of a square-base pyramid is $SA = e^2 + 2es$, where e is an edge of the square base, and s is the slant height of the pyramid.

Cylinders

A *cylinder* differs from prisms and pyramids in that the bases are circles, instead of polygons like squares, rectangles, and triangles. Since the bases are circles, there are no lateral faces. Instead, there is one surface that forms the sides of the cylinder. This surface can be imagined by thinking of a soup can label and how it wraps around a cylindrical soup can. A visual of a cylinder and its net is shown in the figure below.

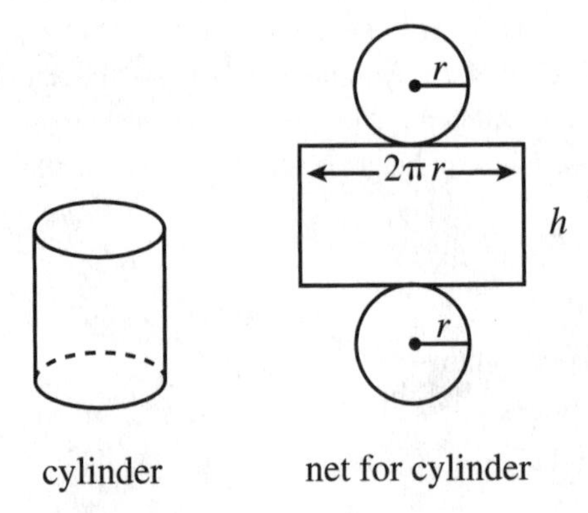

cylinder net for cylinder

The formula for the surface area of a cylinder is based on the net pictured above. Since the bases are two congruent circles, the formula for the area of a circle is used. The area of a circle is $A = \pi r^2$, where r is the radius of the circle. This was explained in Chapter 5. Both bases are circles, so the area of the bases together is equal to $2\pi r^2$, where r is the radius of a base of the cylinder.

The face that was described above as the soup can label is a rectangle. The formula for the area of a rectangle is $A = bh$, where b is the length of the base and h is the height. The base of this rectangle is equal to the circumference of one of the circular bases and the height is equal to the height of the cylinder. Since the formula for the circumference of a circle is $C = 2\pi r$, the area of this rectangular face is equal to $2\pi r \bullet h$, or $2\pi rh$.

Putting all three surfaces together, the formula for the surface area of a cylinder is $SA = 2\pi r^2 + 2\pi rh$, where r is the radius of one of the circular bases and h is the height of the cylinder.

Spheres

A *sphere* is in the shape of a ball. It can be defined as all of the points equidistant from a given point in three dimensions.

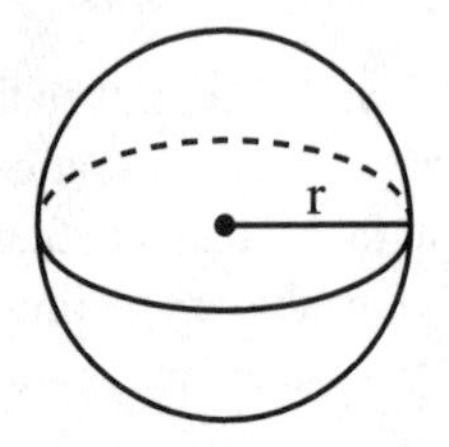

The surface of a sphere is made up entirely of curved surfaces and can be described as the area of four circles. So, the formula for the surface area of a sphere also involves the formula for the area of a circle, as was mentioned above in the formula for the surface area of a cylinder.

Since the surface area can be expressed as the area of four circles, the formula for the surface area of a sphere is $SA = 4\pi r^2$.

COMMON FORMULAS FOR SURFACE AREA

Rectangular prism	$SA = 2lw + 2lh + 2wh$
Cube	$SA = 6e^2$
Square base pyramid	$SA = e^2 + 2es$
Cylinder	$SA = 2\pi r^2 + 2\pi rh$
Sphere	$SA = 4\pi r^2$

STEPS YOU NEED TO REMEMBER

Finding the Surface Area Using Substitution

To find the surface area of any solid figure, you can substitute any known information into the appropriate formula and evaluate. When evaluating, you will be following a procedure known as the order of operations. The correct order of operations is:

Parentheses

Exponents

Multiply and **D**ivide, in order from left to right

Add and **S**ubtract, in order from left to right

This order can be remembered by using the acronym **PEMDAS.**

Take the following example:

What is the surface area of a square-base pyramid with a base edge of 10 meters and a slant height of 4 meters?

To find the surface area, first find the correct formula. The formula for the surface area of a square base pyramid, as stated above, is $SA = e^2 + 2es$, where e is the length of a base edge and s is the slant height of the pyramid.

Next, locate all of the known values for the formula and substitute these values for the variables in the formula. In this problem, the base edge is 10 meters, so $e = 10$. The slant height is 4 meters, so $s = 4$. After substituting, use the correct order of operations to evaluate the formula.

Write the formula: $SA = e^2 + 2es$
Substitute the given values: $SA = 102 + 2(10)(4)$
Evaluate by using the correct order of operations.
Apply the exponent first: $SA = 100 + 2(10)(4)$
Then multiply: $SA = 100 + 80$
Last, add to find the surface area: $SA = 180$ square meters

Although the formula for each figure may be different, the process is similar. First find the formula, substitute the given values, and then evaluate using the correct order of operations.

Finding the Dimension of a Solid when the Surface Area Is Known

There are a number of times when the actual surface area of a figure may be known, but one of the dimensions could be missing. In this case, you will be following the steps of equation solving to find the missing measure. Let's see what this may look like by using the example below.

> A sphere has a surface area of 144π square units. What is the measure of its radius?

The first step in solving for the measure of the radius is to find the appropriate formula to use. The formula for the surface area of a sphere is $SA = 4\pi r^2$. In this case you will also substitute any given values into the formula but will follow the steps of equation solving to solve for r. These steps may seem the opposite to the order of operations in some cases. The surface area of the sphere is equal to 144π square units, so $SA = 144\pi$.

First, write the formula: $SA = 4\pi r^2$
Substitute the given value: $144\pi = 4\pi r^2$
Divide by 4 on each side of the equation to get r^2 alone: $\frac{144\pi}{4\pi} = \frac{4\pi r^2}{4\pi}$
The equation simplifies to: $36 = r^2$
To solve for r, take the square root of each side of the equation: $6 = r$

The radius of the sphere is 6 units.

STEP-BY-STEP ILLUSTRATION OF THE FIVE MOST COMMON QUESTION TYPES

The next section will present you with five of the most common question types associated with surface area. Use the detailed explanations that follow to help guide you in your study.

Question 1: Finding the Surface Area of a Rectangular Prism

What is the surface area of a rectangular prism that has a width of 6 meters, a length of 4 meters, and a height of 2 meters?

(A) 44 m^2

(B) 48 m^2

(C) 88 m^2

(D) 96 m^2

(E) 176 m^2

To find the surface area, first find the correct formula. The formula for the surface area of a rectangular prism is $SA = 2lw + 2lh + 2wh$, where l is the length, w is the width, and h is the height of the prism.

Next, locate all of the known values for the formula and substitute these values in for the variables in the formula. In this problem, the width is 6 meters, so $w = 6$. The length is 4 meters, so $l = 4$. The height is 2 meters, so $h = 2$. The net for this figure is shown in the figure below.

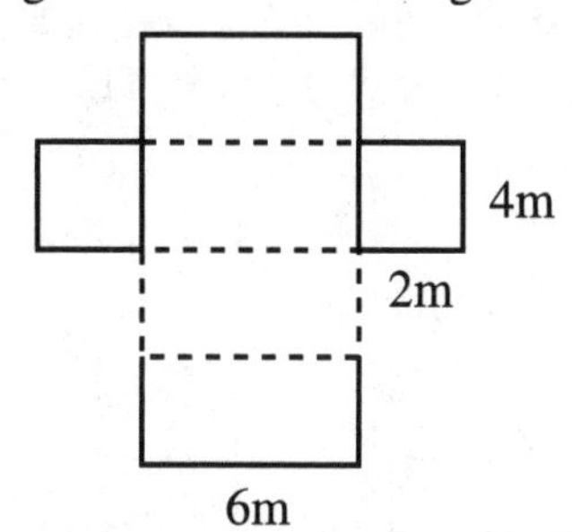

After substituting, use the correct order of operations to evaluate the formula.

Write the formula: $SA = 2lw + 2lh + 2wh$

Substitute the given values: $SA = 2(4)(6) + 2(4)(2) + 2(6)(2)$

Evaluate by using the correct order of operations.

Multiply in each term first: $SA = 2(24) + 2(8) + 2(12)$

Then add to find the surface area: $SA = 48 + 16 + 24$

$SA = 88$ square meters

The correct answer is choice (C). If you selected choice (A), you may have forgotten to multiply by 2 in each term because this value is equal to one-half of the surface area. Choice (B) is the volume of the figure, not the surface area. Choice (D) is the volume doubled. Choice (E) is the surface area doubled, and could be the result of multiplying by two twice.

Question 2: Finding the Length of an Edge of a Cube When the Surface Area Is Known

What is the length of an edge of a cube whose surface area is 150 cm^2?

(A) 3 cm

(B) 5 cm

(C) 10 cm

(D) 15 cm

(E) 25 cm

In this type of problem, the surface area is given and a dimension of the figure is missing. The steps to this process are slightly different from the steps for just finding the surface area. As discussed earlier in the chapter, the known values will be substituted in the formula and the steps for equation solving used to find the missing value.

The first step in solving for the length of an edge is to find the appropriate formula to use. The formula for the surface area of a cube is $SA = 6e^2$, where e represents the length of an edge of the cube. The net for this figure is shown in the figure below.

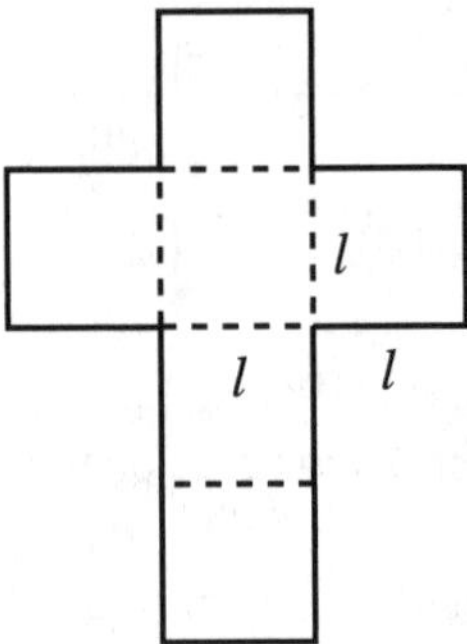

The surface area is 150 cm^2, so $SA = 150$.

First, write the formula: $SA = 6e^2$

Substitute the given values: $150 = 6e^2$

Divide by 6 on each side of the equation to get e^2 alone. $\frac{150}{6} = \frac{6e^2}{6}$

The equation simplifies to: $25 = e^2$

To solve for e, take the square root of each side of the equation. $5 = e$

The length of an edge of the cube is 5 cm.

The correct answer choice is (B). Choices (C) and (D) are the result of dividing values into 150 meters to find factors of 150 and not using the formula for surface area. Choice (E) is the correct value of e^2, but you need to find the value of e by taking the square root in the final step.

Question 3: Finding the Surface Area of a Square-Base Pyramid

What is the surface area of a pyramid with a square base edge of 4 meters and a slant height of 5 meters?

(A) 16 m^2

(B) 20 m^2

(C) 40 m^2

(D) 56 m^2

(E) 96 m^2

To find the surface area of the pyramid, first find the correct formula. The formula for the surface area of a square base pyramid is $SA = e^2 + 2es$, where e is the length of a base edge and s is the slant height of the pyramid.

Next, find all of the known values for the formula and substitute these values in for the variables in the formula. In this problem, the base edge is 4 meters, so $e = 4$. The slant height is 5 meters, so $s = 5$. The net for this figure is shown in the figure below.

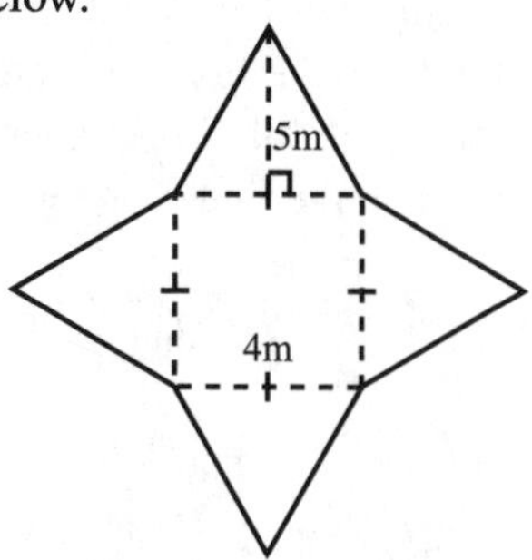

After substituting, use the correct order of operations to evaluate the formula.

Write the formula: $SA = e^2 + 2es$
Substitute the given values: $SA = 4^2 + 2(4)(5)$
Evaluate by using the correct order of operations.
Apply the exponent first: $SA = 16 + 2(4)(5)$
Then multiply: $SA = 16 + 40$
Lastly, add to find the surface area: $SA = 56$ square meters

The correct choice is (D). Choice (A) is the area of the square base of the pyramid, but not the surface area of the entire figure. Choice (B) is the result of multiplying the edge of the square base by the slant height. Choice (C) is the area of the triangular lateral faces of the pyramid, but this value does not include the area of the square base. Choice (E) is the result of not taking one-half of the base • height to find the area of each triangular lateral face.

Question 4: Finding the Surface Area of a Cylinder

What is the surface area, in terms of π, of a cylinder with a height of 12 meters and a base with a radius of 4 meters?

(A) $16\pi m^2$

(B) $32\pi m^2$

(C) $48\pi m^2$

(D) $96\pi m^2$

(E) $128\pi m^2$

To find the surface area of the cylinder, first find the correct formula. The formula for the surface area of a cylinder is $SA = 2\pi r^2 + 2\pi rh$, where r is the radius of the base and h is the height of the cylinder. In this question, all answers are in terms of π, so just leave it in your answer.

Next, find all of the known values for the formula and substitute these values for the variables in the formula. In this problem, the radius of the base is 4 meters, so $r = 4$. The height of the cylinder is 12 meters, so $h = 12$. The net for this figure is shown in the figure below.

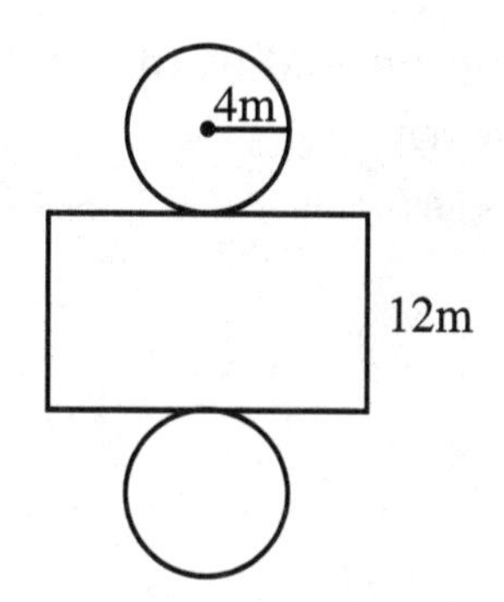

After substituting, use the correct order of operations to evaluate the formula.

Write the formula: $SA = 2\pi r^2 + 2\pi rh$

Substitute the given values: $SA = 2(4)^2 + 2\pi(4)(12)$

Evaluate by using the correct order of operations.

Apply the exponent first: $SA = 2\pi(16) + 2\pi(4)(12)$

Then multiply: $SA = 32\pi + 96\pi$

Lastly, add to find the surface area: $SA = 128\pi$ square meters

The correct choice is (E). Choice (A) is the area of one of the circular bases. Choice (B) is the area of both circular bases. Choice (C) is the result of multiplying the radius of the base by the height of the cylinder and π. Choice (D) is the surface area of the rectangular region that forms the height of the cylinder.

Question 5: Finding the Radius of a Sphere when the Surface Area Is Known

What is the length of the radius of a sphere whose surface area is 196π cm^2?

(A) 7 cm

(B) 14 cm

(C) 49 cm

(D) 98 cm

(E) 784 cm

The first step in solving for the measure of the radius is to find the appropriate formula to use. The formula for the surface area of a sphere is $SA = 4\pi r^2$. In this case you will also substitute any given values into the formula but will follow the steps of equation solving to solve for r. These steps may seem the opposite of order of operations in some cases. The surface area of the sphere is equal to 196π cm^2, so $SA = 196\pi$.

First, write the formula: $SA = 4\pi r^2$

Substitute the given value: $196 = 4\pi r^2$

Divide by 4π on each side of the equation to get r^2 alone. $\frac{196\pi}{4\pi} = \frac{4\pi r^2}{4\pi}$

The equation simplifies to: $49 = r^2$

To solve for r, take the square root of each side of the equation: $7 = r$

The radius of the sphere is 7 units, which is choice (A). Choice (B) is the result of finding the square root of 196, only. Choice (C) is the correct value of r^2, but you need to take the square root in order to find the length of the radius. Choice (D) is the result of dividing 196 by 2. Choice (E) is the result of multiplying 196 by 4 instead of dividing by 4 and taking the square root.

CHAPTER QUIZ

1. What is the surface area of a rectangular prism with a length of 5.5 meters, a width of 7 meters, and a height of 3 meters?

   (A) 76 m^2

   (B) 115.5 m^2

   (C) 150 m^2

   (D) 152 m^2

   (E) 304 m^2

2. What is the surface area of a rectangular prism with a length of 10 inches, a width of 20 inches, and a height of 14 inches?

   (A) 44 in^2

   (B) 620 in^2

   (C) 1,240 in^2

   (D) 2,480 in^2

   (E) 2,800 in^2

3. What is the surface area of a cube whose edge measures 12 cm?

   (A) 72 cm^2

   (B) 144 cm^2

   (C) 432 cm^2

   (D) 864 cm^2

   (E) 1,728 cm^2

4. What is the length of an edge of a cube whose surface area is 486 m^2?

   (A) 9 m

   (B) 54 m

   (C) 81 m

   (D) 121.5 m

   (E) 243 m

5. What is the surface area of a square base pyramid with a base edge of 12 cm and a slant height of 3.5 cm?

   (A) 42 cm^2

   (B) 84 cm^2

   (C) 144 cm^2

   (D) 228 cm^2

   (E) 312 cm^2

6. What is the surface area of a square-base pyramid with a base edge of 20 meters and a slant height of 6 meters?

   (A) 120 m^2

   (B) 240 m^2

   (C) 360 m^2

   (D) 400 m^2

   (E) 640 m^2

7. What is the surface area of a cylinder with a base radius of 7 meters and a height of 3 meters? Leave your answer in terms of π.
   (A) 21π m^2
   (B) 42π m^2
   (C) 49π m^2
   (D) 98π m^2
   (E) 140π m^2

8. What is the height of a cylinder whose surface area is 204π m^2 and has a base radius of 6 meters?
   (A) 11 m
   (B) 16 m
   (C) 23 m
   (D) 34 m
   (E) 1,224 m

9. What is the surface area of a sphere with a radius of length 8 meters? Leave your answer in terms of π.
   (A) 16π m^2
   (B) 64π m^2
   (C) 256π m^2
   (D) 512π m^2
   (E) $2{,}048\pi$ m^2

10. What is the length of the radius of a sphere whose surface area is 784π square units?
   (A) 7 units
   (B) 14 units
   (C) 49 units
   (D) 98 units
   (E) 196 units

ANSWER EXPLANATIONS

1. D

To find the surface area, first find the correct formula. The formula for the surface area of a rectangular prism is $SA = 2lw + 2lh + 2wh$, where l is the length, w is the width, and h is the height of the prism.

Next, locate all of the known values for the formula and substitute these values in for the variables in the formula. In this problem, the length is 5.5 meters, so $l = 5.5$. The width is 7 meters, so $w = 7$. The height is 3 meters so $h = 3$. After substituting, use the correct order of operations to evaluate the formula.

Write the formula: $SA = 2lw + 2lh + 2wh$
Substitute the given values: $SA = 2(5.5)(7) + 2(5.5)(3) + 2(7)(3)$
Evaluate by using the correct order of operations.
Multiply in each term first: $SA = 2(38.5) + 2(16.5) + 2(21)$

Then add to find the surface area: $SA = 77 + 33 + 42$
$SA = 152$ square meters

If you selected choice (A), you may have forgotten to multiply by 2 in each term because this value is equal to one-half of the surface area. Choice (B) is the volume of the figure, not the surface area. Choice (C) is the result of a calculation error. Choice (E) is the surface area doubled, and could be the result of multiplying by two an extra time.

2. C

To find the surface area, first find the correct formula. The formula for the surface area of a rectangular prism is $SA = 2lw + 2lh + 2wh$, where l is the length, w is the width, and h is the height of the prism.

Next, locate all of the known values for the formula and substitute these values in for the variables in the formula. In this problem, the length is 10 inches, so $l = 10$. The width is 20 inches, so $w = 20$. The height is 14 inches so $h = 14$. After substituting, use the correct order of operations to evaluate the formula.

Write the formula: $SA = 2lw + 2lh + 2wh$
Substitute the given values: $SA = 2(10)(20) + 2(10)(14) + 2(20)(14)$
Evaluate by using the correct order of operations.
Multiply in each term first: $SA = 2(200) + 2(140) + 2(280)$
Then add to find the surface area: $SA = 400 + 280 + 560$
$SA = 1{,}240$ square inches

If you selected choice (A), you may have just added the three dimensions together. If you chose (B), you may have forgotten to multiply by 2 in each term because this value is equal to one-half of the surface area. Choice (D) is the surface area doubled and could be the result of multiplying by two an extra time. Choice (E) is the volume of the figure, or the result of simply multiplying the three dimensions together.

3. D

To find the surface area, first find the correct formula. The formula for the surface area of a cube is $SA = 6e^2$, where e is the length of an edge of the cube.

Next, locate the known value for the formula and substitute the value for the variable in the formula. In this problem, the edge is 12 cm, so $e = 12$. After substituting, use the correct order of operations to evaluate the formula.

Write the formula: $SA = 6e^2$
Substitute the given value: $SA = 6(12)^2$
Evaluate by using the correct order of operations.
Apply the exponent: $SA = 6(144)$
Then multiply to find the surface area: $SA = 864 \text{ cm}^2$

Choice (A) is the result of multiplying 6 by 12, and not applying the exponent in the formula. Choice (B) is the surface area of one square face, but not the total surface area of the figure. Choice (C) is equal to one-half of the correct surface area. Choice (E) is the volume of the figure, not the surface area.

4. A

The first step in solving for the length of an edge is to find the appropriate formula to use. The formula for the surface area of a cube is $SA = 6e^2$, where e represents the length of an edge of the cube. The surface area is 486 m^2, so $SA = 486$.

First, write the formula: $SA = 6e^2$
Substitute the given value: $486 = 6e^2$
Divide by 6 on each side of the equation to get e^2 alone. $\frac{486}{6} = \frac{6e^2}{6}$

The equation simplifies to: $81 = e^2$
To solve for e, take the square root of each side of the equation: $9 = e$

The length of an edge of the cube is 9 m.

Choice (B) is the result of multiplying 9 by 6. Choice (C) is the correct value of e^2, but you need to find the value of e by taking the square root in the final step.

Choices (D) and (E) are the result of dividing values into 486 to find factors and not using the formula for surface area.

5. D

To find the surface area of the pyramid, first find the correct formula. The formula for the surface area of a square-base pyramid is $SA = e^2 + 2es$, where e is the length of a base edge and s is the slant height of the pyramid.

Next, find all of the known values for the formula and substitute these values for the variables in the formula. In this problem, the base edge is 12 cm, so $e = 12$. The slant height is 3.5 cm, so $s = 3.5$. After substituting, use the correct order of operations to evaluate the formula.

Write the formula:	$SA = e^2 + 2es$
Substitute the given values:	$SA = 122 + 2(12)(3.5)$
Evaluate b]y using the correct order of operations.	
Apply the exponent first:	$SA = 144 + 2(12)(3.5)$
Then multiply:	$SA = 144 + 84$
Last, add to find the surface area:	$SA = 228$ cm^2

Choice (A) is the result of multiplying the edge of the square base by the slant height. Choice (B) is the area of the triangular lateral faces of the pyramid, but this value does not include the area of the square base. Choice (C) is the area of the square base of the pyramid but not the surface area of the entire figure. Choice (E) is the result of not taking one-half of the base height to find the area of each triangular lateral face.

6. E

To find the surface area of the pyramid, first find the correct formula. The formula for the surface area of a square-base pyramid is $SA = e^2 + 2es$, where e is the length of a base edge and s is the slant height of the pyramid.

Next, find all of the known values for the formula and substitute these values in for the variables in the formula. In this problem, the base edge is 20 meters, so $e = 20$. The slant height is 6 meters, so $s = 6$. After substituting, use the correct order of operations to evaluate the formula.

Write the formula:	$SA = e^2 + 2es$
Substitute the given values:	$SA = 20^2 + 2(20)(6)$
Evaluate by using the correct order of operations.	
Apply the exponent first:	$SA = 400 + 2(20)(6)$
Then multiply:	$SA = 400 + 240$
Last, add to find the surface area:	$SA = 640$ square meters

Choice (A) is the result of multiplying the edge of the square base by the slant height. Choice (B) is the area of the triangular lateral faces of the pyramid, but this value does not include the area of the square base. Choice (C) is the result of a calculation error. Choice (D) is the area of the square base of the pyramid, but not the surface area of the entire figure.

7. E

To find the surface area of the cylinder, first find the correct formula. The formula for the surface area of a cylinder is $SA = 2\pi r^2 + 2\pi rh$, where r is the radius of the base and h is the height of the cylinder. In this question, all answers are in terms of π, so just leave it in your answer.

Next, find all of the known values for the formula and substitute these values in for the variables in the formula. In this problem, the radius of the base is 7 meters, so $r = 7$. The height of the cylinder is 3 meters, so $h = 3$. After substituting, use the correct order of operations to evaluate the formula.

Write the formula:	$SA = 2\pi r^2 + 2\pi rh$
Substitute the given values:	$SA = 2\pi(7)^2 + 2\pi(7)(3)$
Evaluate by using the correct order of operations.	
Apply the exponent first:	$SA = 2\pi(49) + 2\pi(7)(3)$
Then multiply:	$SA = 98\pi + 42\pi$
Last, add to find the surface area:	$SA = 140\pi$ square meters

Choice (A) is the result of multiplying the radius of the base by the height of the cylinder and π. Choice (B) is the surface area of the rectangular region that forms the height of the cylinder. Choice (C) is the area of one of the circular bases. Choice (D) is the area of both circular bases.

8. A

The first step in solving for the measure of the height is to find the appropriate formula to use. The formula for the surface area of a cylinder is $SA = 2\pi r^2 + 2\pi rh$, where r is the radius of the base and h is the height of the cylinder. In this case you will substitute any given values into the formula but will follow the steps of equation solving to solve for h. The surface area of the cylinder is equal to 204π square meters, so $SA = 204\pi$. The base radius is 6 meters, so $r = 6$.

First, write the formula:	$SA = 2\pi r^2 + 2\pi rh$
Substitute the given values:	$204\pi = 2\pi(6)^2 + 2\pi(6)h$
Apply the exponent:	$204\pi = 2\pi(36) + 2\pi(6)h$
Multiply:	$204\pi = 72\pi + 12h$

Subtract 72 from each side of the equation:	$204\pi - 72\pi = 72\pi - 72\pi + 12\pi h$
The equation becomes:	$132\pi = 12\pi h$
Divide by 12 on each side of the equation to get h alone.	$\frac{132\pi}{12\pi} = \frac{12\pi h}{12\pi}$
The value of h is:	$11 = h$

Choice (B) is the result of substituting and not evaluating r^2. Choice (C) is the result of a calculation error. Choice (D) is the result of dividing 204 by the radius of 6. Choice (E) is the result of multiplying 204 by 6.

9. C

To find the surface area of the sphere, first find the correct formula. The formula for the surface area of a sphere is $SA = 4\pi r^2$, where r is the radius of the sphere. In this question, all answers are in terms of π, so just leave it in your answer.

Next, use the known value of the variable in the formula. In this problem, the radius of the sphere is 8 meters, so $r = 8$. After substituting, use the correct order of operations to evaluate the formula.

Write the formula:	$SA = 4\pi r^2$
Substitute the given value:	$SA = 4\pi(8)^2$
Evaluate by using the correct order of operations.	
Apply the exponent first:	$SA = 4\pi(64)$
Then multiply to find the surface area:	$SA = 256\pi$

Choice (A) is the result of multiplying the length of the radius by 2 and π. Choice (B) is the correct value of r^2, but you need to multiply this value by 4π to find the surface area. Choices (D) and (E) are the result of calculation errors.

10. B

The first step in solving for the measure of the radius is to find the appropriate formula to use. The formula for the surface area of a sphere is $SA = 4\pi r^2$. To solve this problem you will need to substitute the given value into the formula and then follow the steps of equation solving to solve for r. The surface area of the sphere is equal to 784π square units, so $SA = 784\pi$.

First, write the formula: $SA = 4\pi r^2$

Substitute the given value: $784\pi = 4\pi r^2$

Divide by 4π on each side of the equation to get r^2 alone. $\frac{784\pi}{4\pi} = \frac{4\pi r^2}{4\pi}$

The equation simplifies to: $196 = r^2$

To solve for r, take the square root of each side of the equation: $14 = r$

Choice (A) is the result of a calculation error. Choice (C) is the result of simply looking for factors of 784. Choice (D) is the result of dividing 196 by 2 instead of taking the square root. Choice (E) is the correct value of r^2, but not the value of r.

CHAPTER 7

Volume

WHAT IS VOLUME?

In the last chapter, you were introduced to three-dimensional figures and surface area. This chapter will review the other major measurement for three-dimensional figures —volume. *Volume* is the number of cubic units it takes to fill a solid. Keep in mind that volume is a three-dimensional calculation, so always use cubic units as a label. The common three-dimensional figures have formulas used to determine the volume.

CONCEPTS TO HELP YOU

This section will help to further your understanding of volume as it relates to various three-dimensional figures. The volume of various types of prisms, the cylinder, the pyramid, the cone, and the sphere will be described.

Volume of Prisms

You were introduced to the concept of prisms in the last chapter on surface area. A *prism* is a three-dimensional figure with congruent, parallel *bases.* These bases can be various shapes, such as a square, a rectangle, a triangle, or any of the other polygons. A prism also has a *height,* which is the length of the edges that connect the parallel bases. Some of the common prisms are shown here:

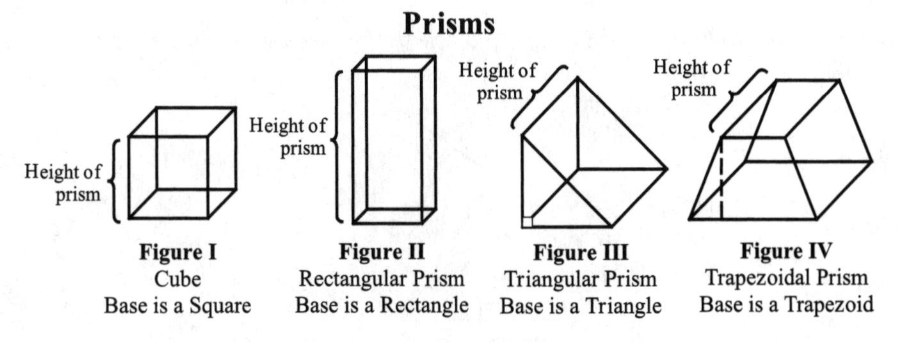

To determine volume, you need to identify the shape of these bases and determine the height of the prism. Notice that the height is not always oriented such that it is vertical. As mentioned previously, volume is the number of cubic units it takes to fill a three-dimensional figure. Look at the prisms on the previous page. Imagine a stack of these bases that are each one unit thick, as shown below for a *rectangular prism* of height 5 units. A rectangular prism has bases that are rectangles.

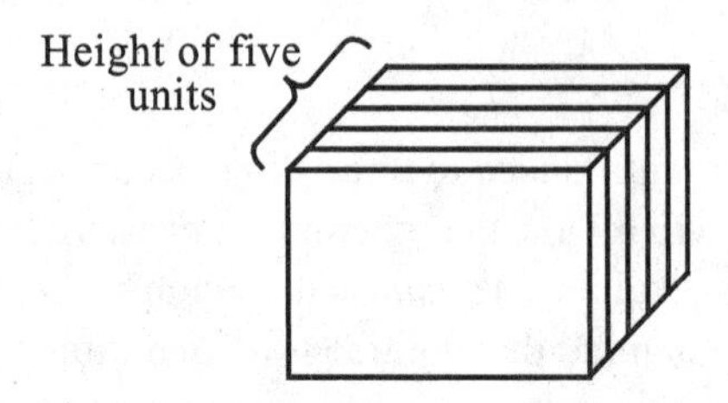

Rectangular Prism

The volume is the area of this rectangular base, multiplied by how many of these are stacked up—in this case, five. That is the concept behind volume, that volume is the area of the base times the height. This can be generally written as $V = Bh$, where B represents the area of the base shape, and h represents the height of the prism. So the area of a rectangular prism is $V = Bh$, where B is the area of the rectangle. The area of a rectangle is found by lw, resulting in $V = lwh$.

A *cube* is a rectangular prism with all edges of equal length. If the edge length is represented as e, the volume is $V = e^3$.

A *triangular prism* has bases that are triangles. For a triangular prism, the area of the base is the formula for the area of a triangle, that is $B = \frac{1}{2}bh$. The volume of a triangular prism is therefore $V = \left(\frac{1}{2}bh_1\right)h_2$, where b is the base of the triangle, h_1 is the height of the triangle, and h_2 is the height of the prism.

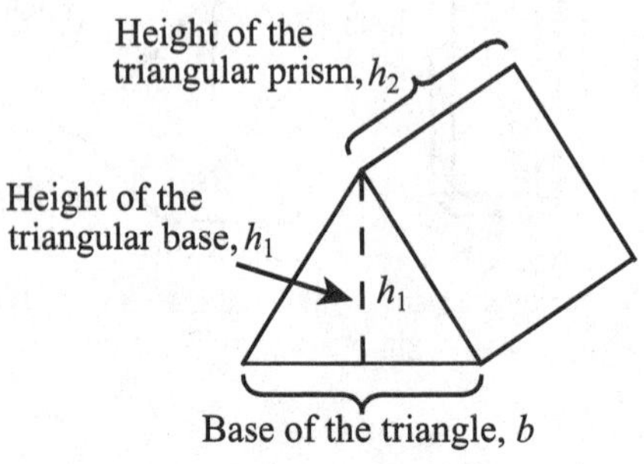

Triangular Prism

Likewise, the volume of a trapezoidal prism is Bh, and B represents the area of the trapezoid. The area of a trapezoid is $\frac{1}{2}h(b_1 + b_2)$ so the volume of a trapezoidal prism is $V = \left(\frac{1}{2}h_1(b_1 + b_2)\right)h_2$, where h_1 is the height of the trapezoid, b_1 and b_2 are the parallel sides of the trapezoid and h_2 is the height of the trapezoidal prism.

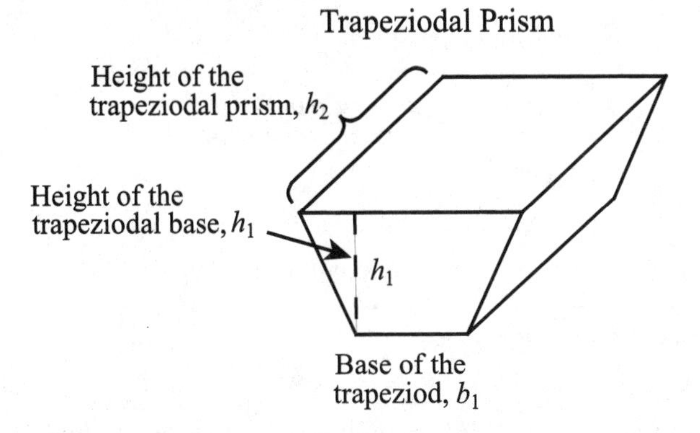

Volume of a Cylinder

A *cylinder* is also a three-dimensional solid but the parallel bases are circles, instead of polygons such as squares, rectangles, or triangles. For volume, the concept is the same, which is that volume is the area of the base times the height. The base of a cylinder is a circle, the area of the base, $B = \pi r^2 h$. Therefore, $V = Bh$ or $V = \pi r^2 h$, where r is the radius of the circular base and h is the height of the cylinder. The symbol π, pronounced "pi," is a constant. Pi is an irrational number that is the ratio of the circumference to the diameter. It is a constant value for every circle and is sometimes rounded to 3.14. Because it is irrational, we generally keep answers in terms of π, such as 36π, 19π, etc., to get a more exact answer. If an answer is required without the constant pi, use the π key on your calculator, not the value 3.14, and round the answer at the end to the required accuracy.

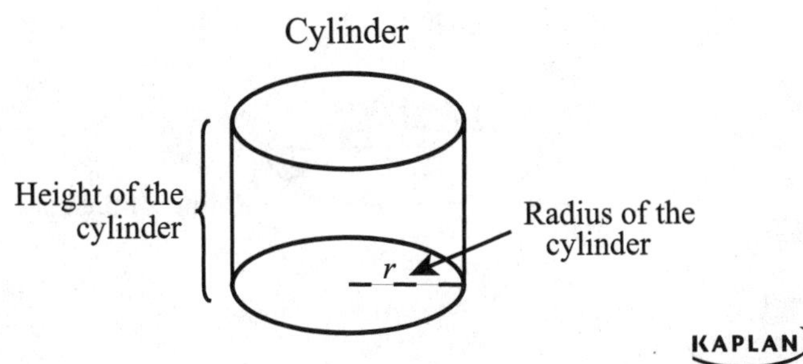

Volume of a Pyramid

A *pyramid* is another of the common three dimensional figures in geometry. Pyramids, like prisms, are defined by the shape of the base. But a pyramid has triangular sides, called *faces*, which meet at a single point. The height of the pyramid itself extends from the single vertex point down to the base and is perpendicular to the base.

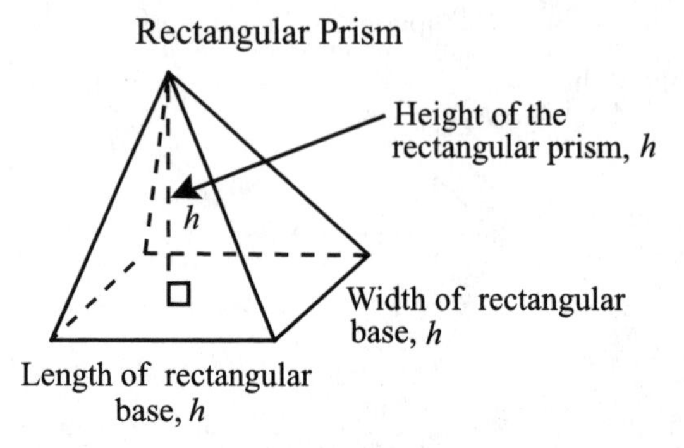

The volume of a pyramid is one-third the volume of the corresponding prism, determined by the shape of the base. The volume of a square pyramid is thus $V = \frac{1}{3}e^3$, and the volume of a rectangular pyramid is $V = \frac{1}{3}lwh$. The height of the pyramid is the length of the perpendicular line segment that extends from the vertex point to the base, as shown above.

Volume of a Cone

The *cone* is a three dimensional figure related to the cylinder. It has a circular base and then all these points meet at a single vertex. The height of the cone is the length of the perpendicular line segment that connects the vertex to the circular base as shown below.

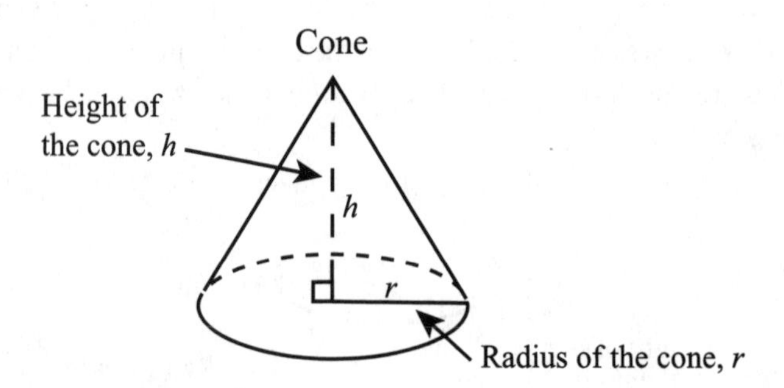

Similar to the pyramid, the cone has a volume that is one-third the volume of its corresponding cylinder. The formula for the volume of a cone is $V = \frac{1}{3}\pi r^2 h$.

Volume of a Sphere

A *sphere* is a three-dimensional object, defined as all the points equidistant from a central point in three dimensions. A sphere has a radius just as a circle does.

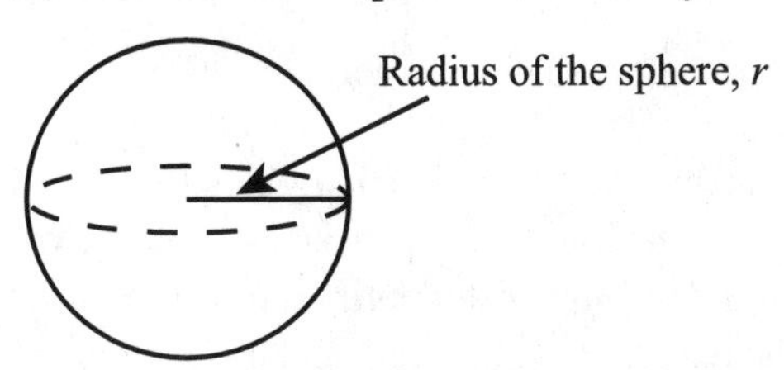

The sphere does not have a base like the rest of the three-dimensional solids discussed. To find the volume of a sphere, use the formula $V = \frac{4}{3}\pi r^3$.

COMMON FORMULAS FOR VOLUME

Rectangular Prism	$V = lwh$
Cube	$V = e^3$
Cylinder	$V = \pi r^2 h$
Pyramid	$V = \frac{1}{3}lwh$
Cone	$V = \frac{1}{3}\pi r^2 h$
Sphere	$V = \frac{4}{3}\pi r^3$

STEPS YOU NEED TO REMEMBER

This section will take you through the steps needed to use the common formulas to determine volume. You will also be instructed on how to find a missing measurement if the volume is known, by using algebra.

Finding Volume Using Substitution in the Formula

To find the volume of one of the common three-dimensional figures, you can substitute all the given information into the appropriate formula and evaluate. When evaluating, you must remember to follow the correct order of operations. Recall that the correct order of operations is:

Parentheses

Exponents

Multiply and **D**ivide, in order from left to right

Add and **S**ubtract, in order from left to right

This order can be remembered by the acronym **PEMDAS.**

For example, a rectangular prism has a length of 18 mm, a width of 12 mm and a height of 8 mm, and you are asked to find the volume. You are given the measurements, now just identify the correct formula, that is $V = lwh$, where l is the length, w is the width and h is the height. Substitute the measures for the unknowns in the formula and you have $V = (18)(12)(8) =$ 1,728 mm^3.

Here's another example:

Find the volume of a cylinder with a base diameter of 14 inches and a height of 20 inches. Use the correct formula, $V = \pi r^2 h$, and substitute in the correct values for r, the radius, and h, the height. You are not told the radius; you are given the diameter. Recall that the radius is one-half of the diameter, so the radius is $14 \div 2 = 7$ inches. The height, h, is given as 20 inches. After substituting, use the correct order of operations to evaluate.

Write the formula:	$V = \pi r^2 h$
Substitute the correct values:	$V = \pi \bullet 7^2 \bullet 20$
Now, using the correct order of operations, evaluate the exponent first:	$V = \pi \bullet 49 \bullet 20$
Then multiply and add the cubic units:	$V = 980\,\pi \bullet in^3$

The volume of the cylinder is 980π in^3.

A trapezoidal prism has a number of measures that are used in the formula

$$V = \left(\frac{1}{2}h_1(b_1 + b_2)\right)h_2$$

To find the volume of the following prism, first correctly identify the needed measures.

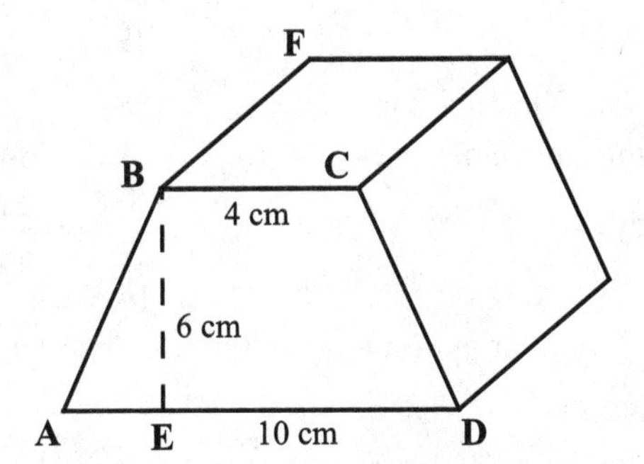

The height of the trapezoid base, h_1, is 6 cm. The parallel sides of the trapezoid base, b_1 and b_2, are 4 cm and 10 cm. The height of the prism, h_2, is 12 cm.

Write the formula: $V = \left(\frac{1}{2}h_1(b_1 + b_2)\right)h_2$

Substitute the correct values: $V = \frac{1}{2} \bullet 6 \bullet (4+10) \bullet 12$

Now, use the correct order of operations and simplify parentheses first, then multiply: $V = \frac{1}{2} \bullet 6 \bullet 14 \bullet 12$

$\text{V} = 504 \text{ cm}^3$

The volume of the trapezoidal prism is 504 cm^3.

Finding the Dimension of a Solid when the Volume is Known

There are occasions when the volume of a three-dimensional figure is known, but one of the measures is missing. In this case, you will set up the formula and substitute in the given measures. Let a variable represent the unknown and then solve the resultant equation to find the missing measure. Consider the example below:

> A rectangular pyramid has a volume of 196 mm^3. If the length is 7 mm and the height is 8 mm, what is the width of the pyramid?

The formula for the volume of a rectangular pyramid is $V = \frac{1}{3}lwh$. The volume, V, is 196 mm^3, the length, l, is 7 mm and the height, h, is 8 mm. You can represent the width s the variable w.
Write the correct formula: $V = \frac{1}{3}lwh$

Substitute the given values:	$196 = \frac{1}{3} \bullet 7 \bullet w \bullet 8$
Multiply both sides by 3 to clear the fraction:	$588 = 7 \bullet 8 \bullet w$
Multiply on the right-hand side:	$588 = 56w$
Divide both sides by 56:	$\frac{588}{56} = \frac{56w}{56}$
Simplify:	$10.5 = w$

The width of the rectangular pyramid is 10.5 mm.

Here is another example:

> A cone has a volume of 384π m^3. What is the radius of the circular base, if the height is 18 m?

The first step to solve for the measure of the radius is to find the appropriate formula. This formula is $V = \frac{1}{3}\pi r^2 h$. Let r represent the unknown radius measure. The volume, V, is given as $384\pi m^3$, and the height, h, is given as 18m. Substitute the given values into the formula and solve for the radius, r.

The formula is:	$V = \frac{1}{3}\pi r^2 h$
By substitution, the equation is:	$384\pi = \frac{1}{3} \bullet \pi \bullet r^2 \bullet 18$
Multiply on the right-hand side:	$384\pi = 6\pi r^2$
Divide both sides by 6π:	$\frac{384\pi}{6\pi} = \frac{6\pi r^2}{6\pi}$
Simplify:	$64 = r^2$
Take the square root of both sides:	$\sqrt{64} = \sqrt{r^2}$
The radius is 8m.	$8 = r$

Another type of question you may encounter will give the volume and some of the dimensions in terms of a variable as in the figure below.

> If the volume of the prism below is 320 in^3, what is the value of x?

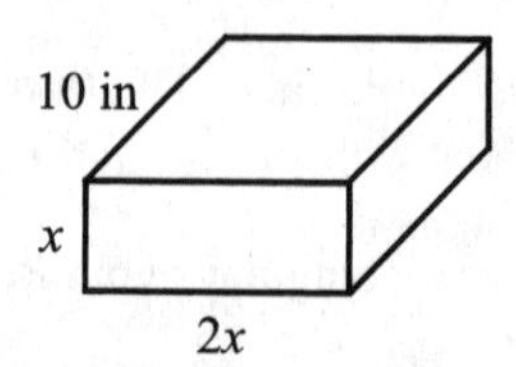

Find the appropriate formula, in this case $V = lwh$. The volume, V, is 320 in^3, the length, l, is represented as $2x$, the width, w, is represented as x and the height, h, is 10 inches. Substitute in the given values and solve for x:

Write the formula:	$V = lwh$
Substitute in the values to get the equation:	$320 = x(2x)(10)$
Multiply on the right hand side:	$320 = 20x^2$
Divide both sides by 20:	$\frac{320}{20} = \frac{20x^2}{20}$
Simplify:	$16 = x^2$
Take the square root of both sides:	$\sqrt{16} = \sqrt{x^2}$
The value of x is 4 inches.	$4 = x$

The formula for each figure is different, but you will use the same process to find a missing dimension when the volume is given.

STEP-BY-STEP ILLUSTRATION OF THE FIVE MOST COMMON QUESTION TYPES

Question 1: Finding the Volume of a Prism

Find the volume of the figure below:

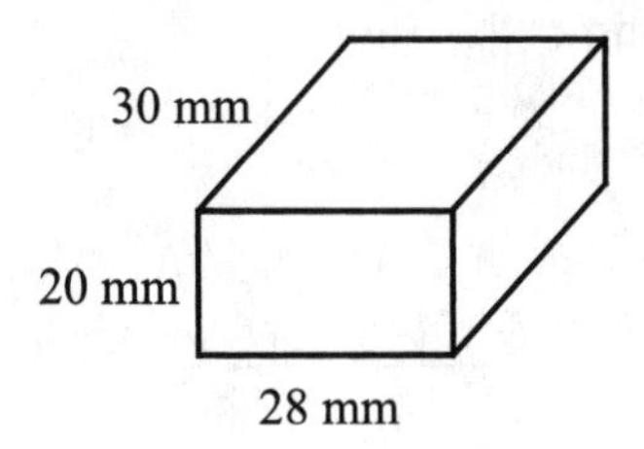

(A) 78 mm^3

(B) 560 mm^3

(C) 590 mm^3

(D) 4,000 mm^3

(E) 16,800 mm^3

This is a rectangular prism, and the formula for the volume is $V = lwh$. In this solid, the length, l, is 28 mm, the width, w, is 20 mm and the height, h, is 30 mm. Substitute in the given dimensions and evaluate. $V = (28)(20)(30) = 16{,}800$ mm^3, **the correct answer is (E).**

If you added the dimensions instead of multiplying, you would have arrived at incorrect answer choice (A). Choice (B) is the area of the base rectangle, not the volume. You may have chosen (C) if you had taken the area of the base rectangle and added in the height, instead of multiplying. Choice (D) represents the surface area of the prism, not the volume.

Question 2: Finding a Missing Dimension when the Volume is Given

Find the diameter of a cylinder if the volume is $1{,}014\pi$ in^3 and the height is 6 inches.

(A) 13 inches

(B) 26 inches

(C) 84.5 inches

(D) 169 inches

(E) 338 inches

The formula for the volume of a cylinder is $V = \pi r^2 h$. The volume, V, is $1{,}014\pi$ cubic inches and the height, h, is 6 inches. Substitute the given measures in the formula and then solve the equation for r. Once you determine the radius, r, the diameter is two times this value.

Write the formula: $V = \pi r^2 h$

Substitute in the given values: $1{,}014\pi = \pi r2(6)$

Divide both sides by 6π: $\frac{1014\pi}{6\pi} = \frac{\pi r^2 6}{6\pi}$

Simplify: $169 = r^2$

Take the square root of both sides: $\sqrt{169} = \sqrt{r^2}$

$13 = r$

The radius is 13 inches, so the diameter is $13 \bullet 2 = 26$ inches. **The correct answer is choice (B).** Answer choice (A) is the value of the radius, not the diameter. If your choice was answer (C), you divided 169 by 2 instead of taking the square root. Choice (D) is the value of the radius squared. Choice (E) is two times the radius squared.

Question 3: Finding the Volume of a Pyramid

Find the volume of the pyramid:

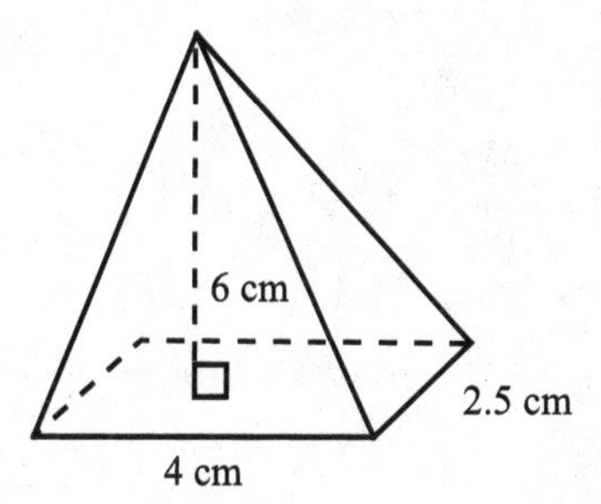

(A) 4.17 cm^3

(B) 10 cm^3

(C) 12.5 cm^3

(D) 20 cm^3

(E) 60 cm^3

This is a rectangular pyramid, so the formula for the volume is $V = \frac{1}{3}lwh$. The length, *l*, is 4 cm, the width, *w*, is 2.5 cm and the height, *h*, is 6 cm. Substitute the given dimensions to get $V = \frac{1}{3} \bullet 4 \bullet 2.5 \bullet 6$. Multiply all terms on the right-hand side, and the volume is 20 cm^3, **so the correct answer is choice (D).**

If you had added the three dimensions and then taken one-third of this sum, you would have chosen answer (A). Choice (B) is the area of the base rectangle. Choice (C) is just the sum of the three dimensions. If you chose (E), you gave the volume of a rectangular prism; you forgot to take one-third of the product of the length, width, and height.

Question 4: Finding the Height of a Cone When the Volume Is Known

What is the height of a cone if the volume is 168π mm^3 and the radius is 6 mm?

(A) 4.7 mm

(B) 9.3 mm

(C) 14 mm

(D) 42 mm

(E) 84 mm

The formula for the volume of a cone is $V = \frac{1}{3}\pi r^2 h$. The volume, V, is given as 168π mm^3, and the radius, r, is 6 mm. Substitute in the correct values and then solve for h, the height.

Write the formula: $V = \frac{1}{3}\pi r^2 h$

Substitute in the values given to get the equation: $168\pi = \frac{1}{3}\pi 6^2 h$

Use order of operations and evaluate the exponent: $168\pi = \frac{1}{3}\pi 36 \bullet h$

Multiply on the right-hand side: $168\pi = 12\pi h$

Divide both sides by 12π: $\frac{168\pi}{12\pi} = \frac{12\pi h}{12\pi}$

$14 = h$

The height is 14 mm, **the correct answer choice of (C).** If you chose (A) you forgot the one-third factor in the formula, as if it was a cylinder instead of a cone. Choice (B) would be incorrectly arrived at if you had used one-half instead of one-third in the formula. Choice (D) would be the result of incorrectly evaluating the 6^2 term as $6 \bullet 2$. If you chose (E), you may have just used r instead of r^2 in the formula.

Question 5: Finding the Volume of a Sphere

What is the volume of a sphere with a radius of 3 inches?

(A) 9π in^3

(B) 12π in^3

(C) 27π in^3

(D) 36 in^3

(E) 36π in^3

The formula for the volume of a sphere is $V = \frac{4}{3}\pi r^3$. Substitute in the value for the radius, $r = 3$ inches, and evaluate.

The formula is: $V = \frac{4}{3}\pi r^3$

Substitute in the value for r to get: $V = \frac{4}{3}\pi 3^3$

Use order of operations and evaluate the exponent: $V = \frac{4}{3}\pi \bullet 27$

Simplify: $V = 36\pi$ in^3

The correct answer is choice (E). If your answer choice was (A), you used one-third instead of four-thirds in the formula. If you chose (B), you evaluated 3^3 as 3 • 3 instead of 3 • 3 • 3. If your answer was (C) you forgot the four-thirds factor in the formula. Answer choice (D) would be arrived at if you forgot the π factor.

CHAPTER QUIZ

1. What is the volume of a cube with sides that are 5 cm in length?

   (A) 15 cm^3

   (B) 25 cm^3

   (C) 30 cm^3

   (D) 125 cm^3

   (E) 150 cm^3

2. If the volume of the rectangular prism below is 1,960 m^3, what is the value of *x*?

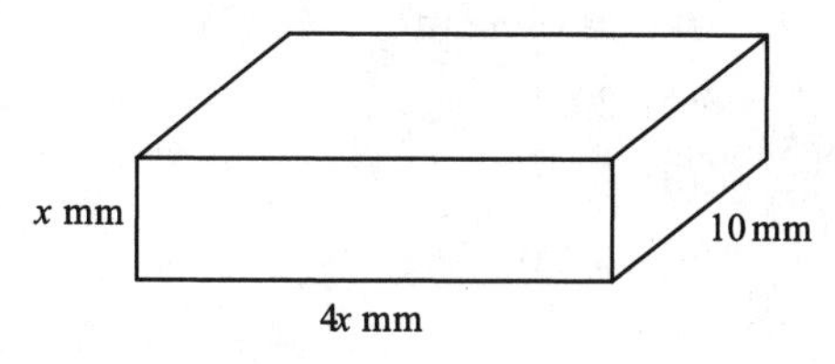

   (A) 7 m

   (B) 14 m

   (C) 22.14 m

   (D) 49 m

   (E) 390 m

3. Find the volume of the trapezoidal prism below:

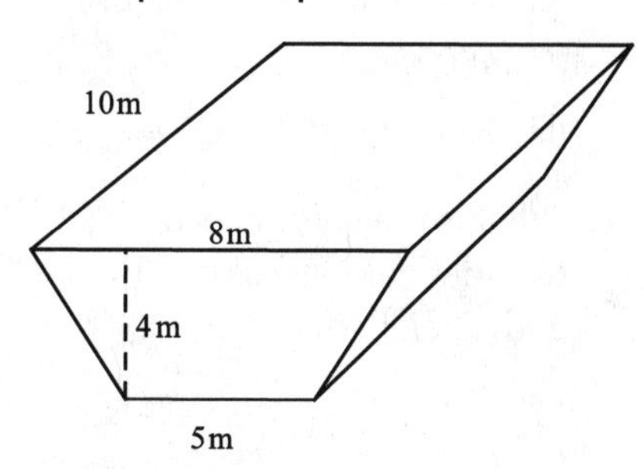

   (A) 26 in^3

   (B) 260 in^3

   (C) 320 in^3

   (D) 520 in^3

   (E) 800 in^3

4. If the volume of the triangular prism below is 36 cm^3, what is the measure of the height, *h*, of the prism?

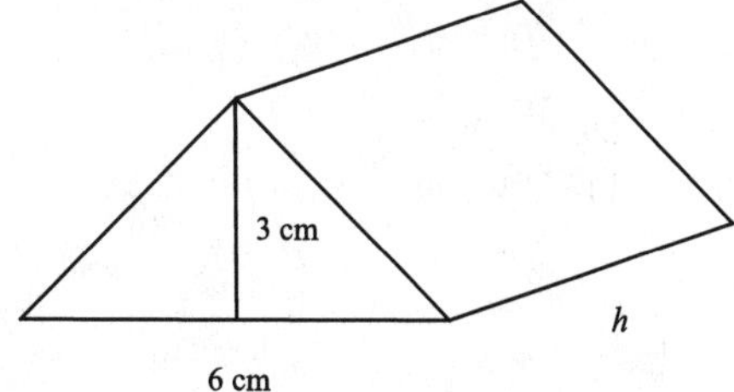

   (A) 1 cm

   (B) 2 cm

   (C) 4 cm

   (D) 9 cm

   (E) 27 cm

5. What is the volume of a rectangular pyramid that has base dimensions of 8 mm and 12 mm and a height of 5 mm?
   (A) 25 mm^3
   (B) 160 mm^3
   (C) 240 mm^3
   (D) 480 mm^3
   (E) 1,440 mm^3

6. What is the volume of a cylinder with radius of 11 cm and a height of 4 cm?
   (A) 44π cm^3
   (B) 88π cm^3
   (C) 161.3π cm^3
   (D) 484π cm^3
   (E) 968π cm^3

7. If the volume of a cylinder is 600π mm^3, and the height is 6 mm, what is the diameter of the cylinder?
   (A) 5 mm
   (B) 10 mm
   (C) 20 mm
   (D) 50 mm
   (E) 100 mm

8. What is the volume of the cone shown below?

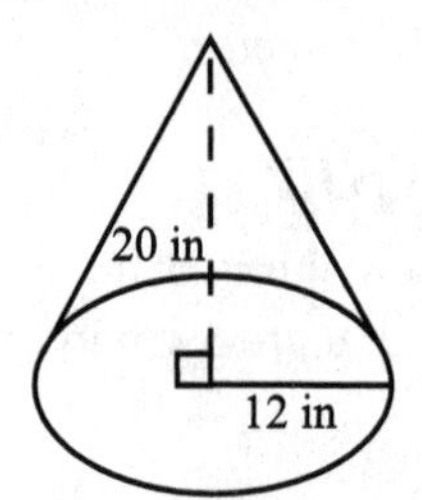

   (A) 80π in
   (B) 240π in^3
   (C) 960π in^3
   (D) $2{,}880\pi$ in^3
   (E) $8{,}640\pi$ in^3

9. What is the height of a cone with volume of 24π cm^3 and a radius of 3 cm?
   (A) 2.67 cm
   (B) 3 cm
   (C) 6 cm
   (D) 8 cm
   (E) 12 cm

10. What is the volume of a sphere with a diameter of 16 mm?
   (A) 32π mm^3
   (B) 170.67π mm^3
   (C) 256π mm^3
   (D) 512π mm^3
   (E) 682.67π mm^3

ANSWER EXPLANATIONS

1. D

A cube is a rectangular prism, where all sides are the same length. The formula for the volume of a cube is $V = e^3$, where e is the length of the side, which in this problem is 5 cm. Substitute in this value to get $V = 53$, or $V = 5 \bullet 5 \bullet 5 = 125 \text{ cm}^3$. If you chose answer (A), you multiplied $5 \bullet 3$ instead of $5 \bullet 5 \bullet 5$. Answer choice (B) is the area of the base of the prism. If you chose (C), you multiplied the number of faces of the prism, 6, by the side length. Choice (E) is the surface area of the prism, not the volume.

2. A

To answer this problem, first determine the correct formula. The formula for the volume of a rectangular prism is $V = lwh$. The volume, V, is given as 1,960 m^3, the length, l, is represented as $4x$, the width, w, is represented as x, and the height, h, is given as 10 mm. Substitute into the formula and then solve for the variable.

Write the formula:	$V = lwh$
Substitute in the values:	$1960 = 4x \bullet x \bullet 10$
Multiply the terms on the right hand side:	$1960 = 40x^2$
Divide both sides by 40:	$\frac{1960}{40} = \frac{40x^2}{40}$
Simplify:	$49 = x^2$
Take the square root of both sides:	$\sqrt{49} = \sqrt{x^2}$
Simplify:	$7 = x$

If your answer was choice (B), you omitted the coefficient of 4 and had the incorrect equation of $1960 = 10x^2$. Choice (C) would be incorrectly arrived at if you forgot to factor in the height in the formula, and had the equation $1960 = 4x^2$. Choice (D) is the value of x^2, not the value of x. If your answer was choice (E), you may have added the dimensions, instead of multiplying, to get the incorrect equation $1960 = 5x + 10$.

3. B

The volume of a trapezoidal prism is found by using the formula

$$V = \left(\frac{1}{2}h_1(b_1 + b_2)\right)h_2 .$$

Use the figure to correctly identify the height of the trapezoid base, h_1, as 4 m, the parallel sides of the trapezoid base, b_1 and b_2, as 5m and 8m, and the height of the prism, h_2, as 10 m. Substitute into the formula to get $V = \frac{1}{2} \bullet 4 \bullet (5+8) \bullet 10$ Use the correct order of operations and evaluate the parentheses first: $V = \frac{1}{2} \bullet 4 \bullet 13 \bullet 10$. Multiply all terms, and the volume is 260 m^3. Choice (A) is the area of the trapezoidal base, not the volume of the prism. Choice (C) would be the volume of a rectangular prism of dimensions 8 m, 4 m, and 10 m. If your answer was choice (D), you forgot the one-half term in the formula. Choice (E) would be incorrectly arrived at by multiplying the parallel sides, b^1 and b^2, instead of adding them.

4. C

First, identify the correct formula for a triangular prism, which is $V = \left(\frac{1}{2} bh_1\right) h_2$. According to the figure, the height of the triangle base, h_1, is 3 cm, the base of the triangle, b, is 6 cm, and the height of the prism, h_2, is represented as h.

Write the formula:	$V = \left(\frac{1}{2} bh_1\right) h_2$
Substitute in the given values:	$36 = \frac{1}{2} \bullet 6 \bullet 3 \bullet h$
Multiply on the right-hand side:	$36 = 9h$
Divide both sides by 9:	$\frac{36}{9} = \frac{9h}{9}$
Simplify:	$4 = h$

The height is 4 cm.

If your answer was choice (A), you used an incorrect formula, such as $V = (2bh_1)h_2$. If your answer was (B) you used the incorrect formula $V = bh_1h_2$. Choice (D) could be the result of just adding the two given dimensions, 3 + 6 = 9. Choice (E) would be incorrectly arrived at if you thought that volume was equal to $b + h_1 + h_2$.

5. B

Use the correct formula for the volume of a rectangular pyramid: $V = \frac{1}{3} lwh$. Substitute in the given values of the length, l, as 12 mm, the width, w, as 8 mm and the height, h, as 5 mm. So the volume is $V = \frac{1}{3} \bullet 12 \bullet 8 \bullet 5 = 160$ mm^3. If you chose (A), you added the three given dimensions. Choice (C) would be arrived at if you used the incorrect formula with a factor of one-half instead of one-third. Choice (D) is the volume of a rectangular prism, not a rectangular pyramid. Choice (E) would be arrived at if you used a factor of 3 in the formula, instead of one-third.

6. D

Use the formula for the volume of a cylinder, $V = \pi r^2 h$. Substitute in the values of the radius, r, as 11 cm and the height, h, as 4 cm. The volume is $V = \pi \bullet 112 \bullet 4$. Use correct order of operations and evaluate the exponent first, $V = \pi \bullet 121 \bullet 4$. Multiply to get $V = 484\pi$ cm^3. If your answer was choice (A), you did not square the radius. Answer choice (B) would be arrived at if you multiplied $11 \bullet 2$ instead of $11 \bullet 11$. Choice (C) is the volume of a cone with these dimensions. Choice (E) is two times the volume of the cylinder.

7. C

Use the formula for the volume of a cylinder, that is $V = \pi r^2 h$. It is given that the volume, V, is 600π mm^3, and the height, h, is 6 mm. Use the formula to solve for the radius, r, and then determine the diameter as twice the radius.

Write the formula:	$V = \pi r^2 h$
Substitute in the values:	$600\pi = \pi \bullet r^2 \bullet 6$
Divide both sides by 6π:	$\frac{600\pi}{6\pi} = \frac{6\pi r^2}{6\pi}$
Simplify:	$100 = r^2$
Take the square root of both sides:	$\sqrt{100} = \sqrt{r^2}$
Simplify:	$10 = r$

The radius is 10 mm, so the diameter is 20 mm. Choice (A) is one-half the value of the radius. Choice (B) is the length of the radius. Choice (D) is an incorrect value of the radius, if you had divided 100 by 2 when solving the equation, instead of taking the square root. Choice (E) is the incorrect value of the diameter, if you had divided 100 by 2 when solving the equation, instead of taking the square root.

8. C

The formula for the volume of a cone is $V = \frac{1}{3}\pi r^2 h$. In this figure, the radius, r, is 12 inches and the height of the cone, h, is 20 inches. Substitute

these values into the formula to get $V = \frac{1}{3}\pi r^2 h = \frac{1}{3} \bullet \pi \bullet 12^2 \bullet 20$.
Use the correct order of operations and evaluate the exponent first:
$V = \frac{1}{3} \bullet \pi \bullet 144 \bullet 20$. Multiply all terms and the volume is 960π in^3. If your answer was choice (A), you forgot that the radius is squared. Choice (B) would be the result of using the incorrect formula $V = \pi rh$. Choice (D) is the volume of a cylinder with these dimensions. Choice (E) would be incorrectly arrived at if you had used a factor of 3, instead of one-third, in the formula.

9. D

To find the height, use the formula for the volume of a cone, that is $V = \frac{1}{3}\pi r^2 h$. The volume, V, is given as 24π cm^3, and the radius, r, is given as 3 cm.

Write the formula:	$V = \frac{1}{3}\pi r^2 h$
Substitute in the given values:	$24\pi = \frac{1}{3} \bullet \pi \bullet 3^2 \bullet h$
Use the correct order of operations and evaluate the exponent:	$24\pi = \frac{1}{3} \bullet \pi \bullet 9 \bullet h$
Multiply on the right-hand side:	$24\pi = 3\pi h$
Divide both sides by 3π:	$\frac{24\pi}{3\pi} = \frac{3\pi h}{3\pi}$
Simplify:	$8 = h$

The height is 8 cm. If your answer was choice (A), you found the height of a cylinder with this volume. Choice (B) is the value of the radius, not the height. Choice (C) is the value of the diameter of the cone. Choice (E) would be incorrectly arrived at if you had multiplied the radius by 2, instead of squaring the radius when using order of operations.

10. E

To find the volume of a sphere, use the formula $V = \frac{4}{3}\pi r^3$. The radius, r, is not given; you are told that the diameter is 16 mm. The radius is one half the value of the diameter, or $r = 8$ mm. Substitute into the formula and

$V = \frac{4}{3} \bullet \pi \bullet 8^3$. Use the correct order of operations and evaluate the exponent first to get $V = \frac{4}{3} \bullet \pi \bullet 512$. Now multiply, and the volume is 682.67π mm^3.

If your answer was (A), you evaluated the formula as $V = \frac{4}{3} \bullet \pi \bullet r \bullet 3$, instead of $V = \frac{4}{3} \bullet \pi \bullet r \bullet r \bullet r$. In choice (B), you may have used a factor of one-third in the formula instead of four-thirds. Choice (C) is the surface area of the sphere. If your answer was (D), you forgot the four-thirds factor in the formula.

CHAPTER 8

Coordinate Geometry

WHAT IS COORDINATE GEOMETRY?

Coordinate geometry is the study of geometric figures on the coordinate, or Cartesian, plane. This plane is made up of a system of axes, called the x-axis and the y-axis. These two signed number lines are perpendicular and intersect at a location called the *origin.* These axes determine the location of all points in the plane. The concept of the coordinate plane has many real-world applications such as longitude and latitude. There are even city streets that were constructed using the principles of coordinate geometry. In this chapter you will review the formulas and equations that are used to help solve problems involving a coordinate system.

CONCEPTS TO HELP YOU

This section will review the various facts and information pertaining to coordinate geometry and the basic process of plotting points in a coordinate system.

Setting Up the Coordinate System

The setup of the coordinate system starts with two intersecting number lines. One is a horizontal line known as the *x-axis* and the other is a vertical line known as the *y-axis.* Since these two lines are perpendicular, they meet to form right angles. When the two lines intersect, four sections, or *quadrants*, are formed. Starting at the upper right hand section, these quadrants are numbered I, II, III, and IV in a counterclockwise fashion, as shown in the figure below. The point of intersection of the two axes is known as the *origin* and has the coordinates (0, 0).

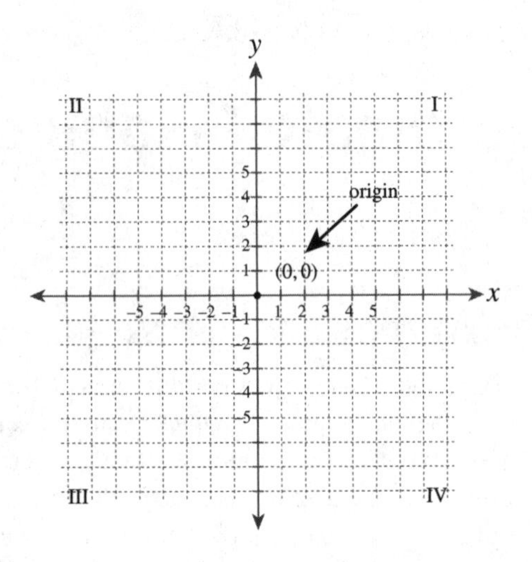

Plotting Points in the Coordinate System

The location of every point in the coordinate plane has an ordered pair associated with it. An *ordered pair* contains an x-value, or x-coordinate, and a y-value, or y-coordinate. These coordinates are determined by the horizontal and vertical distance a point is from the origin. Each value is then written within a set of parentheses in the form (x, y).

In the coordinate system, moving to the right and up from the origin are positive directions. In addition, moving to the left and down from the origin are negative directions. Therefore, any point in quadrant I has positive values for both x and y. In quadrant II, the x-coordinates are negative while the y-coordinates are positive. In quadrant III, both x- and y-coordinates are negative. In quadrant IV, the x-coordinates are positive while the y-coordinates are negative.

Take the following example of how to plot a point. Start with the point A(2, 3). This point has a positive x-coordinate and a positive y-coordinate. To find the location of the point, begin at the origin. Since the x-coordinate is positive, move two units to the right of the origin. Since the y-coordinate is also positive, move three units up from there to find the location of the point. This point is located in quadrant I and is shown in the figure on the next page.

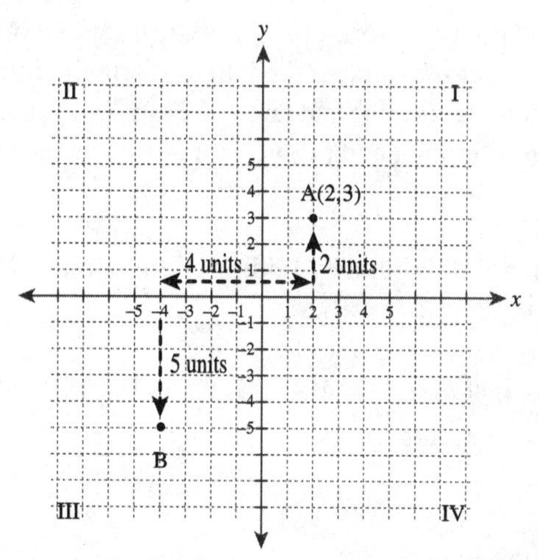

For another example, find the coordinates of point B from the figure above. This point is located 4 four units to the left of the origin and 5 units down. Therefore, the coordinates of point B are (–4, –5). This point is located in quadrant III.

STEPS YOU NEED TO REMEMBER

This section will cover the most common formulas used in coordinate geometry and how to execute them when solving problems in a coordinate system.

Using the Midpoint Formula

The midpoint formula is an important one in the study of coordinate geometry. The *midpoint* of a line segment divides the segment into two congruent parts. This means that each part has the same length. The formula will help you find the midpoint of a line segment in a coordinate system, or possibly an endpoint of a segment if the midpoint and other endpoint are known.

MIDPOINT FORMULA

The formula to find the midpoint of a segment is *midpoint* =

$$\left(\frac{x_1 + x_2}{2}, \frac{y_1 + y_2}{2}\right).$$

In order to use the midpoint formula to find the midpoint of a segment, take the endpoints and substitute them into the formula. In the formula, the endpoints are written as (x_1, y_1) and (x_2, y_2). Thus, (x_1, y_1) represents the coordinates of one endpoint and (x_2, y_2) represents the coordinates of the other endpoint.

Try using the midpoint formula by finding the midpoint of the line segment with endpoints (–1, 3) and (5, –9).

Start by using the point $(x_1, y_1 1, 3)$ as (x_1, y_1) and (5, –9) as (x_2, y_2).

Substitute these values into the midpoint formula: $\left(\frac{x_1 + x_2}{2}, \frac{y_1 + y_2}{2}\right) = \left(\frac{-1+5}{2}, \frac{3+-9}{2}\right)$

Combine the values in the numerators: $\left(\frac{4}{2}, \frac{-6}{2}\right)$

Simplify each fraction to find the coordinates: (2, –3)

The midpoint of the line segment with endpoints (–1, 3) and (5, –9) is (2, –3). The figure below shows the two points and the midpoint on a coordinate graph.

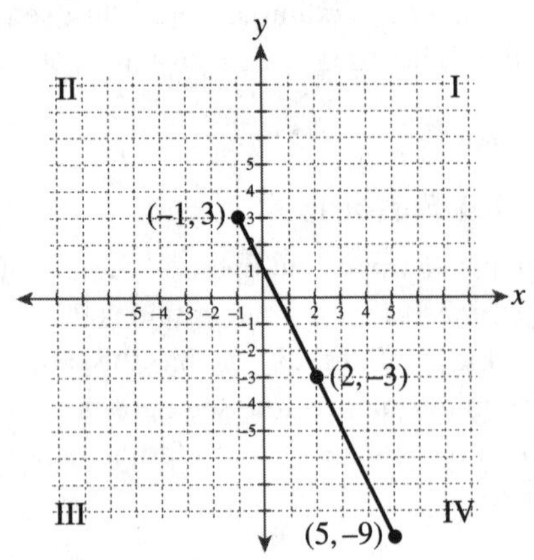

Using the Distance Formula

The distance between two points in a coordinate system can be easy to calculate. For example, the distance between the points (4, 2) and (7, 2) is 3 units as seen on the figure below. The line containing the points is a horizontal line, so the distance can be found by simply counting the units.

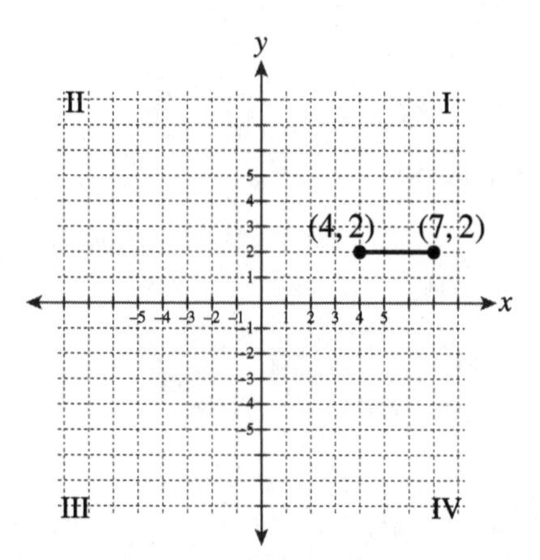

However, how is the distance found if the points are not located on a vertical or horizontal line? In this case, the distance formula can be used along with the coordinates of the endpoints to find the length of the line segment that connects them.

DISTANCE FORMULA

The distance formula is $d = \sqrt{(x_1 - x_2)^2 + (y_1 - y_2)^2}$

It is derived from the Pythagorean theorem, a relationship between the three sides of any right triangle. You can read more about this important theorem in Chapter 12.

Let's practice using this formula with the question below.

What is the distance between the points (–2, 3) and (4, 7)?

The first step is to use the distance formula by substituting the values from the two points. Use the point (–2, 3) as (x_1, y_1) and (4, 7) as (x_2, y_2).

Substitute the values into the formula in order to find the distance between the points:

$$d = \sqrt{(x_1 - x_2)^2 + (y_1 - y_2)^2}$$

Evaluate within the parentheses: $d = \sqrt{(-2-4)^2 + (3-7)^2}$

Apply the exponent: $d = \sqrt{(-6)^2 + (-4)^2}$

Add under the square root symbol: $d = \sqrt{36+16}$

$d = \sqrt{52}$

The distance between the points is units. $\sqrt{52} \approx 7.2$

The figure below shows the two points and the distance between them.

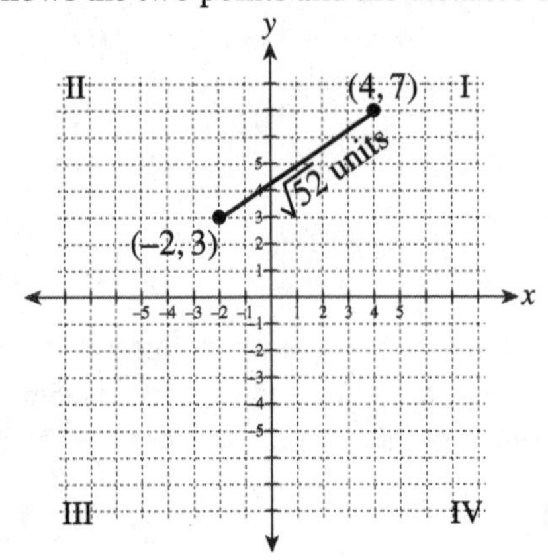

Using the Slope Formula

The *slope* of the line between any two points in the coordinate plane is the difference between the *y*-values divided by the difference in the *x*-values. This is the measure of the steepness of a line. This relationship can be expressed by the formula, where (x_1, y_1) represents the coordinates of one point and (x_2, y_2) represents the coordinates of the other point. The letter m is often used to represent the slope of a line.

THE SLOPE OF A LINEAR EQUATION

Slope is the change in *y* over the change in *x*, and is often referred to as $\frac{rise}{run}$.

The formula for slope is $m = \frac{y_1 - y_2}{x_1 - x_2}$.

A line with positive slope slants up to the right. A line with negative slope slants up to the left.

In order to find the slope between any two points, substitute the values for each coordinate into the formula and simplify. Find, for example, the slope of the line between the points (9, –1) and (5, 7). The first step is to take the formula for the slope of the line and substitute the values.

Use the points (9, –1) as (x_1, y_1) and (5, 7) as (x_2, y_2).

$$m = \frac{y_1 - y_2}{x_1 - x_2}$$

Substitute into the formula: $m = \frac{-1-7}{9-5}$

Simplify the numerator and denominator: $m = \frac{-8}{4}$

Divide to simplify the fraction: $m = -2$

The slope of the line between the two points is –2.

The figure below shows the graph of the two points and the slope between them.

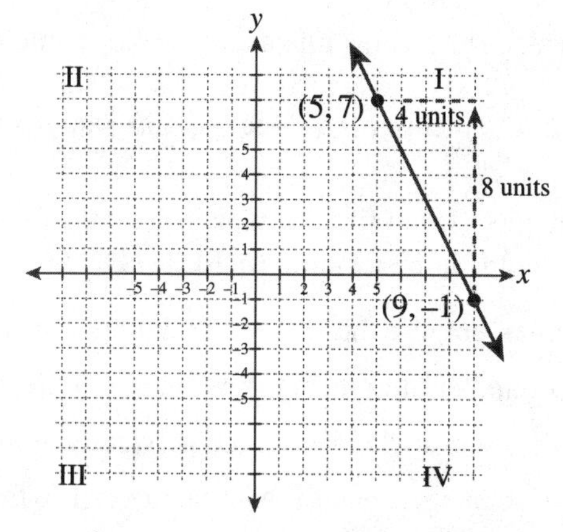

Slope–Intercept Form of a Linear Equation

The most common form of a linear equation is the slope–intercept form. The slope-intercept form is $y = mx + b$, where m represents the slope of the line and b represents the y-intercept. The slope of a line was explained in the previous section. The y-intercept is the place on the graph where the line crosses the y-axis. This form of the equation allows you to graph the equation by using those two facts: the slope a nd the y-intercept.

First, practice identifying the slope and y-intercept in a few equations. Keep in mind that the slope of the line is the coefficient, or number in front of x. The y-intercept will be the term that is added or subtracted to the x term.

1. $y = 5x + 3$ $m = 5$ and $b = 3$
2. $y = \frac{2}{3}x - 4$ $m = \frac{2}{3}$ and $b = -4$

3. $y - 6 = -2x$ $m = -2$ and $b = 6$

In this case, you first need to change the equation to the slope-intercept form by solving for y.

To do this, add six to each side of the equation: $y - 6 + 6 = -2x + 6$
The equation is now: $y = -2x + 6$
The coefficient of x is -2, so $m = -2$ and $b = 6$.

The slope–intercept form of a linear equation can be very handy when graphing linear equations on a set of coordinate axes. After identifying the slope and y-intercept, these numbers are used to create the graph of a straight line.

Use the following steps to graph a line using the slope-intercept form:

1. The first step is to identify the slope (m) and y-intercept (b) in the equation.
2. Use the y-intercept (b) to find the place where the line crosses the y-axis. Make a point at the location $(0, b)$.
3. From the y-intercept, use the slope to find more points on the line. Count up the number of units in the numerator of the slope and over the number of places in the denominator. Make a point at this location. If the slope is an integer, place the integer over 1 to form a fraction (i.e. $4 = \frac{4}{1}$)
4. Continue this process of using the slope until you have a few points on your graph. Connect the points with a straight edge and label the line with the equation.

POSITIVE AND NEGATIVE SLOPES

To graph a line with a positive slope, count up and over to the right.
To graph a line with a negative slope, count up and over to the left.

Practice graphing linear equations using the examples below.

Graph the equation $y = \frac{1}{2}x + 3$.

First, identify the slope (m) and the y-intercept (b). The slope is $\frac{1}{2}$ and the y-intercept is 3.

Graph the y-intercept on the graph. Since $b = 3$, graph the point (0, 3).

Use the slope to count up and over from the y-intercept. Since the slope is $\frac{1}{2}$, count up one unit and over two units to the right to find another point on the line.

Continue to count one unit up and over two units to the right to find a few other points on the line. This equation has been graphed correctly below.

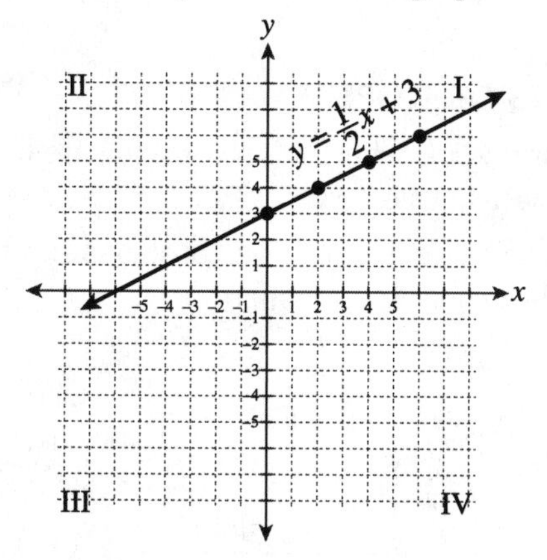

Graph the equation $y = -\frac{3}{2}x - 1$.

1. First, identify the slope (m) and the y-intercept (b). The slope is $-\frac{3}{2}$ and the y-intercept is -1.
2. Graph the y-intercept on the graph. Since $b = -1$, graph the point (0, -1).
3. Use the slope to count up and over from the y-intercept. Since the slope is $\frac{1}{2}$, count up three units and over two units to the left to find another point on the line.
4. Continue to count three units up and over two units to the left to find a few other points on the line. This equation has been graphed correctly below.

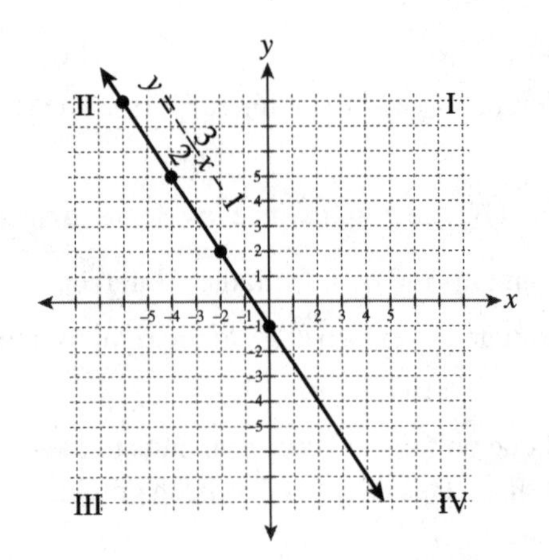

SPECIAL TYPES OF SLOPE

Lines in the form $y = k$, where k is a constant, are horizontal lines. The slope of these lines is zero.

Lines in the form $x = k$, where k is a constant are vertical lines. The slope of these lines is undefined.

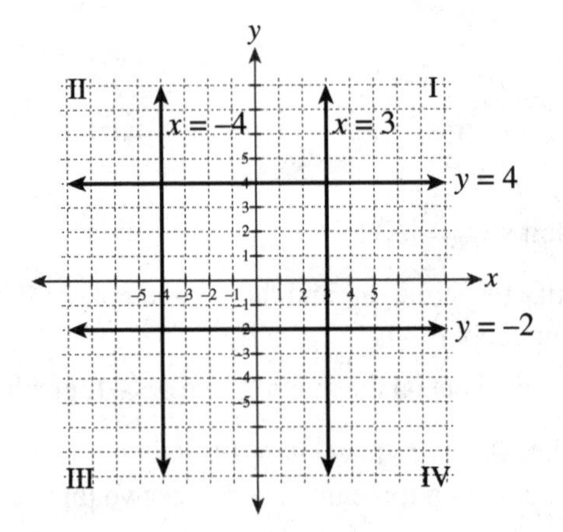

STEP-BY-STEP ILLUSTRATION OF THE FIVE MOST COMMON QUESTION TYPES

Question 1: Plotting Points in a Coordinate System

What are the coordinates of point P in the figure below?

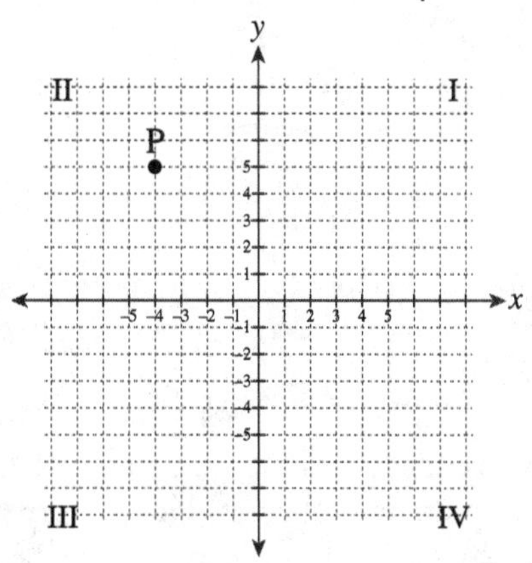

(A) (4, –5)

(B) (5, –4)

(C) (–5, 4)

(D) (–4, –5)

(E) (–4, 5)

The correct answer is choice (E). The location of points in the coordinate plane is based on their horizontal and vertical distance from the origin, or the point (0, 0). To find the coordinates of point P, start at the point (0, 0). Move horizontally four units to the left of the origin on the x-axis to get to a place directly below the point. This is the value $x = -4$. From there, move five units vertically (straight up) and parallel to the y-axis to get to point P. This is the value y = 5. Therefore, the coordinates of the point are (–4, 5). This point is located in quadrant II.

Choice (A) is a point located in quadrant IV that is four units to the right of the origin and five units down from the x-axis. Choice (B) is also located in quadrant IV and is five units to the right of the origin and four units down from the x-axis. Choice (C) is located in quadrant II but is five units to the left of the origin and four units up from the x-axis. Choice (D) is four units to the left of the origin and five units down from the x-axis. It is located in quadrant III.

Question 2: Finding an Endpoint of a Line Segment When the Midpoint is Given

A line segment has a midpoint of (2, 4) and an endpoint of (–3, 6). What are the coordinates of the other endpoint of this segment?

(A) (–0.5, 5)

(B) (7, 2)

(C) (2, 7)

(D) (5, –0.5)

(E) (1, 2)

The correct choice is (B). In order to find the coordinates of the endpoint, use the midpoint formula. The formula to find the midpoint between two points is *midpoint* = $\left(\frac{x_1 + x_2}{2}, \frac{y_1 + y_2}{2}\right)$. However, this question is a little different because the missing part is an endpoint, not the midpoint itself. Since you know one endpoint and the midpoint, substitute into the formula the information about the known endpoint and set each part equal to the midpoint. Treat the known endpoint as (x_1, y_1) and the coordinates of the midpoint as $(x_{midpoint}, y_{midpoint})$.

The equations are: $\frac{x_1 + x_2}{2} = x_{midpoint}$ and $\frac{y_1 + y_2}{2} = y_{midpoint}$

Substitute the given values: $\frac{-3 + x_2}{2} = 2$ and $\frac{6 + y_2}{2} = 4$

Cross-multiply in each proportion: $-3 + x_2 = 4$ and $6 + y_2 = 8$

Solve for each value: $x_2 = 7$ and $y_2 = 2$

The missing endpoint is (7, 2). Choice (A) is the midpoint of the two given points as the endpoints of the line segment and choice (D) is this midpoint with the coordinates reversed. Choice (C) is the result of confusing the x and y coordinates. Choice (E) is the result of an error in the last step of solving the equation for x_2. You subtracted 3 from 4 to get 1 instead of adding 3 to 4 to get 7.

Question 3: Using the Distance Formula

What is the distance between the points (6, 9) and (3, 11)?

(A) $\sqrt{5}$

(B) $\sqrt{13}$

(C) 5

(D) 13

(E) $\sqrt{481}$

The distance between any two points in the coordinate plane can be found by using the distance formula. The distance formula is $d = \sqrt{(x_1 - x_2)^2 + (y_1 - y_2)^2}$.

In order to use this formula, take the two points (6, 9) and (3, 11) and treat them as (x_1, y_1) and (x_2, y_2), respectively.

Substitute the values into the formula in order to find the distance between the points.

$$d = \sqrt{(x_1 - x_2)^2 + (y_1 - y_2)^2}$$

Evaluate within the parentheses: $d = \sqrt{(6-3)^2 + (9-11)^2}$

Apply the exponents: $d = \sqrt{(3)^2 + (-2)^2}$

Add under the square root symbol: $d = \sqrt{9+4}$

$$d = \sqrt{13}$$

The distance between the points is $\sqrt{13} \approx 3.6$ units.

The correct choice is (B). Choice (A) is the result of subtracting the two values under the square root symbol. If you chose answer (C), you may have subtracted the two values under the radical sign after they were squared and omitted the square root symbol. Choice (D) would be the result of omitting the square root symbol. Choice (E) is the result of adding the values of x_1 and x_2, and the values of y_1 and y_2, instead of subtracting them as stated in the formula.

Question 4: Finding the Slope of a Line between Two Points

What is the slope of the line that passes through the points (–3, 4) and (1, –2)?

(A) –3

(B) $-\frac{3}{2}$

(C) $-\frac{2}{3}$

(D) $\frac{2}{3}$

(E) $\frac{3}{2}$

In order to find the slope between any two points, substitute the values for each coordinate into the formula and simplify.

The first step is to take the formula for the slope of the line and substitute the values.

Use the point (–3, 4) as (x_1, y_1) and (1, –2) as (x_2, y_2).

$$m = \frac{y_1 - y_2}{x_1 - x_2}$$

Substitute into the formula: $m = \frac{4-(-2)}{-3-1}$

Simplify the numerator and denominator: $m = \frac{6}{-4}$

Simplify the fraction: $m = -\frac{3}{2}$

The slope of the line between the two points is $-\frac{3}{2}$, **which is choice (B).** Choice (A) may have been chosen if you thought the formula for slope was $\frac{x_1}{x_2}$. Choice (C) is the reciprocal of the correct answer. Choice (D) is the opposite of the reciprocal of the correct choice. Either (C) or (D) may have been chosen if the x-values were placed in the numerator, instead of the denominator and the y-values were placed in the denominator. If you selected choice (E), you found a positive value and the slope of this line is negative.

Question 5: Using the Slope–Intercept Form of a Linear Equation

What is the equation of the line in the figure below?

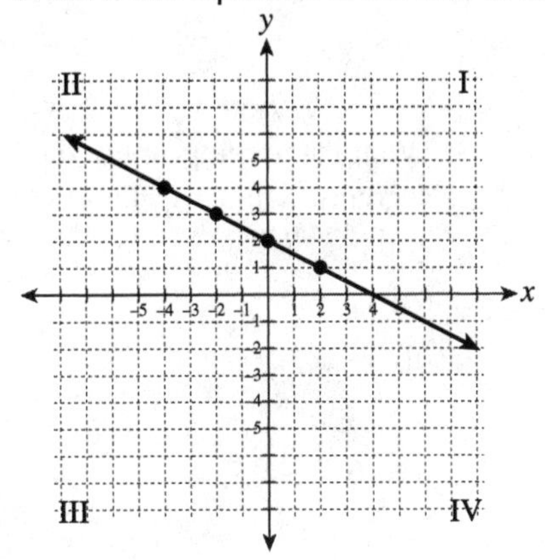

(A) $y=\frac{1}{2}x+2$

(B) $y=-\frac{1}{2}x-2$

(C) $y=2x+2$

(D) $y=-2x+2$

(E) $y=-\frac{1}{2}x+2$

The correct answer is (E). The first step to identify the equation of the line is to find the *y*-intercept. This is the location where the line crosses the *y*-axis. The point on the graph where this line crosses the *y*-axis is (0, 2), so the *y*-intercept is 2. This is the value of b in the equation $y = mx + b$.

The next step is to find the slope of the line. To do this, start at the *y*-intercept and count up and over to get to the next point on this line. The next point on the line from the *y*-intercept is up one unit and over two units to the left. A slope that is up one unit and over two units is a slope of $\frac{1}{2}$. Since the slope was found by counting up and over to the left, the slope is equal to $-\frac{1}{2}$. This is the value of m in the equation $y = mx + b$.

To find the equation of the line, substitute the values for m and b into the slope-intercept form of the equation. The equation is $y=-\frac{1}{2}x+2$.

Choice (A) has the correct y-intercept, but the slope is positive. Choice (B) has the correct slope, but has a y-intercept of –2. Choice (C) has a slope of 2 and a y-intercept of 2. Choice (D) has a slope of –2 and a y-intercept of 2.

CHAPTER QUIZ

1. Which point in the figure below is plotted at the coordinates (–2, 5)?

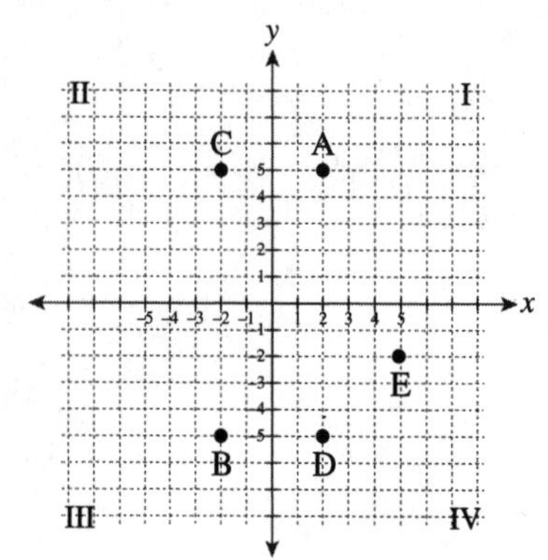

(A) A

(B) B

(C) C

(D) D

(E) E

2. In what quadrant is the point (3, –4) located in?

(A) I

(B) II

(C) III

(D) IV

(E) V

3. What is the midpoint of the line segment with endpoints (3, 10) and (–5, 2)?

(A) (1, 6)

(B) (–2, 12)

(C) (4, 6)

(D) (6.5, –1.5)

(E) (–1, 6)

4. Line segment $\overline{CD}$ contains endpoint C (–4, –5) and midpoint (2, 1). What are the coordinates of endpoint D?

(A) (8, 7)

(B) (–1, –2)

(C) (–2, –1)

(D) (7, 8)

(E) (0, –3)

5. What is the distance between the points (9, –2) and (3, 5)?

(A) $\sqrt{13}$

(B) $\sqrt{85}$

(C) $\sqrt{153}$

(D) 13

(E) 85

6. What is the length of line segment $\overline{RS}$ in the figure below?

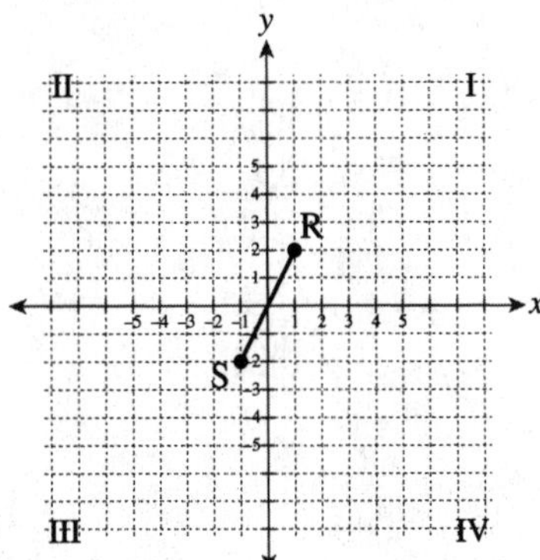

(A) $\sqrt{2}$
(B) $\sqrt{12}$
(C) $\sqrt{20}$
(D) 5
(E) 20

7. What is the slope of the line containing the two points (8, −4) and (−13, 10)?

(A) −3
(B) $-\frac{3}{2}$
(C) $-\frac{2}{3}$
(D) $\frac{2}{3}$
(E) $\frac{3}{2}$

8. What is the slope of the line graphed in the figure below?

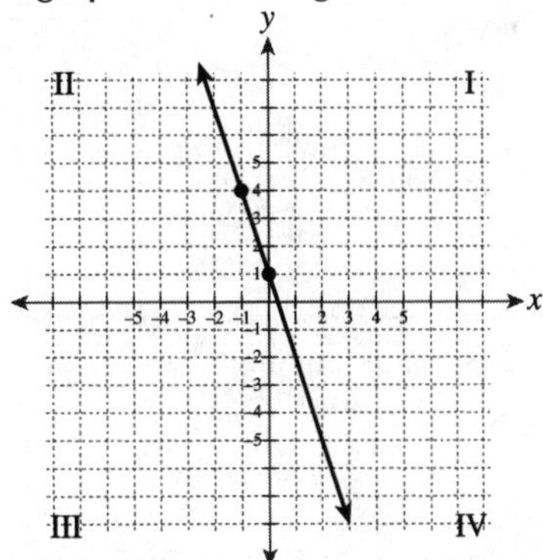

(A) $-\frac{1}{3}$
(B) $\frac{1}{3}$
(C) 1
(D) −3
(E) 3

9. What is the *y*-intercept of the line in the figure below?

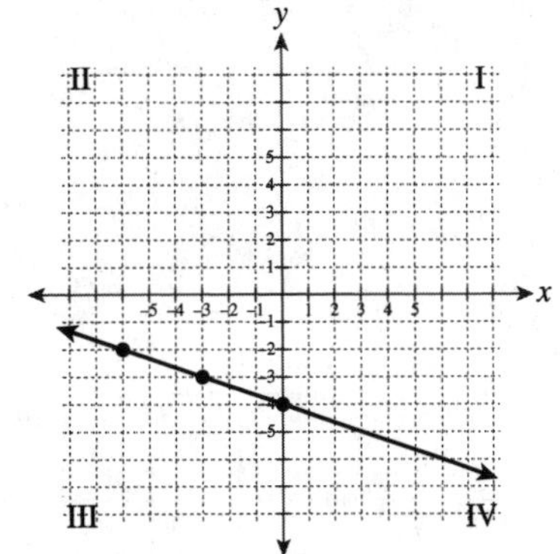

(A) $\frac{1}{3}$
(B) $-\frac{1}{3}$
(C) $\frac{1}{4}$
(D) −4
(E) 4

10. Which of the following is the graph of the equation $2y - 2 = 6x$?

(A)

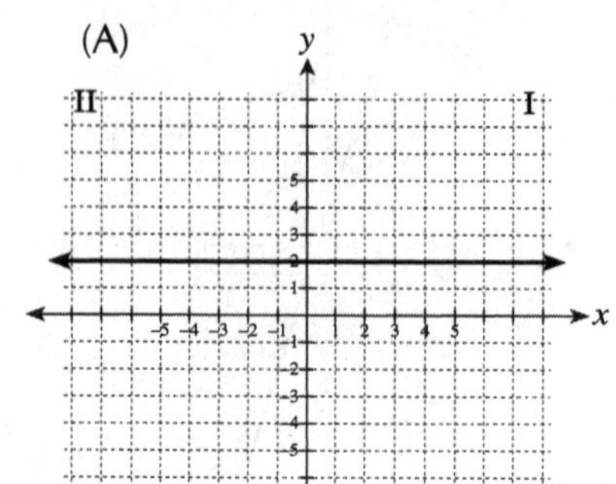

(B)

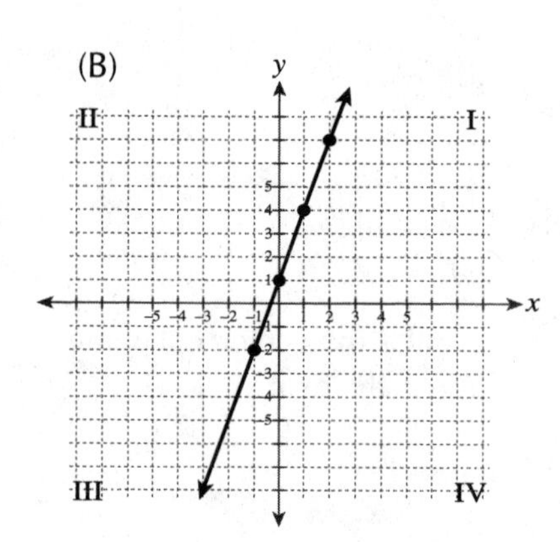

(C)

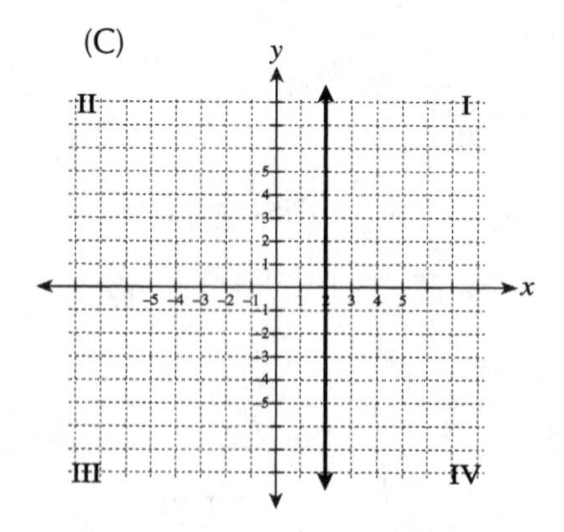

(D)

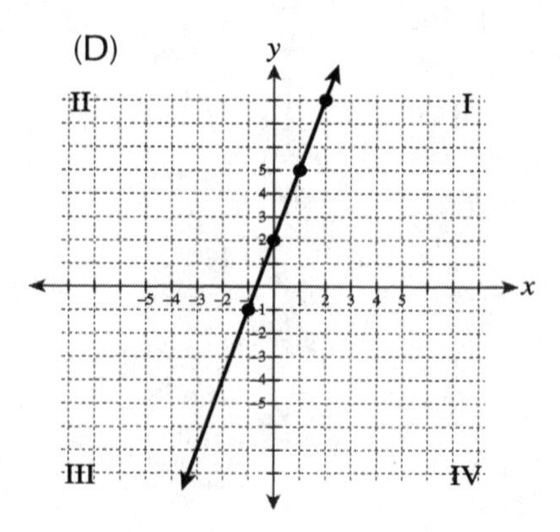

(E)

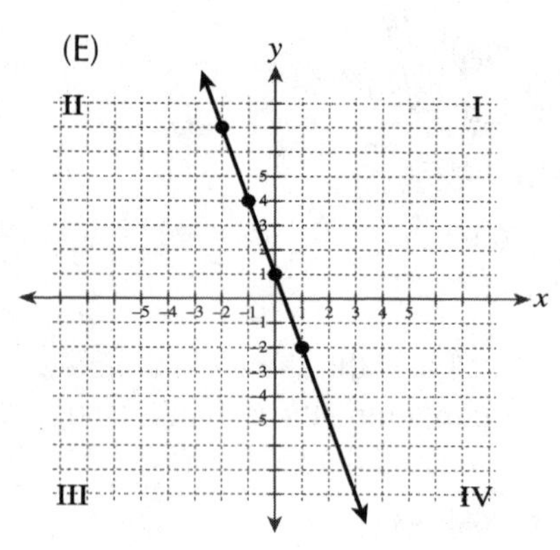

ANSWER EXPLANATIONS

1. C

Each point is located using two values; an x-coordinate and a y-coordinate. To find the location of a point, start at the origin. Take the x-coordinate and move to the right if the value is positive and to the left if it is negative. Then, use the y-coordinate and move up if the value is positive and down if the value is negative. The point (–2, 5) has an x-coordinate of –2, so count over 2 units to the left. Since the y-coordinate is 5, count 5 units up from there. This is the location of point C on the graph.

Point A is at the location (2, 5). Point B is at the location (–2, –5). Point D is at the location (2, –5). Point E is located at (5, –2).

2. D

The coordinate system consists of two intersecting number lines, known as the x- and y-axes. When these two perpendicular lines intersect, four sections, or quadrants, are formed. The quadrants are numbered I, II, III, and IV in a counterclockwise fashion starting with the upper right–hand section. The point (3, –4) is found by starting at the origin and counting 3 units to the right and 4 units down. This location is in quadrant IV.

Choice (B) may have been selected if the location of the point had been found incorrectly by counting up three units and over four units to the left. Choice (E) should have been immediately eliminated because there are only four quadrants.

3. E

Start by using the point (3, 10) as (x_1, y_1) and (–5, 2) as (x_2, y_2).

Substitute these values into the midpoint formula: $\left(\frac{x_1 + x_2}{2}, \frac{y_1 + y_2}{2}\right) = \left(\frac{3 + -5}{2}, \frac{10 + 2}{2}\right)$

Combine the values in the numerators: $\left(\frac{-2}{2}, \frac{12}{2}\right)$

Simplify each fraction to find the coordinates: (–1, 6)

The midpoint of the line segment with endpoints (3, 10) and (–5, 2) is (–1, 6).

Choice (A) may have been selected if the negative sign was left out of the *x*-coordinate. Choice (B) is the result of not dividing each sum by two. Choice (C) is the result of finding the average *x* and *y* values and disregarding the signs of the numbers. Choice (D) is the result of finding the average of the coordinates of the first point as the *x*-coordinate of the midpoint, and finding the average of the coordinates of the second point as the *y*-coordinate of the midpoint.

4. A

In order to find the coordinates of the endpoint, use the midpoint formula. The formula to find the midpoint between two points is *midpoint* = $\left(\frac{x_1 + x_2}{2}, \frac{y_1 + y_2}{2}\right)$. However, this question is a little different because the missing part is an endpoint, not the midpoint itself. Since you know one endpoint and the midpoint, substitute into the formula the information about the known endpoint and set each part equal to the midpoint. Treat the known endpoint as (x_1, y_1) and the coordinates of the midpoint as $(x_{midpoint}, y_{midpoint})$.

The equations are: $\frac{x_1 + x_2}{2} = x_{midpoint}$ and $\frac{y_1 + y_2}{2} = y_{midpoint}$

Substitute the given values: $\frac{-4 + x_2}{2} = 2$ and $\frac{-5 + y_2}{2} = 1$

Cross–multiply in each proportion: $-4 + x_2 = 4$ and $-5 + y_2 = 2$

Solve for each value: $x_2 = 8$ and $y_2 = 7$

The missing endpoint is (8, 7). Choice (B) is the midpoint of the two given points as the endpoints of the line segment and choice (C) is this incorrect midpoint with the coordinates reversed. Choice (D) is the result of confusing the *x* and y coordinates. Choice (E) is the result of an error in the last step of solving the equations for x_2 and y_2.

5. B

The distance between any two points in the coordinate plane can be found by using the distance formula. The distance formula is

$d = \sqrt{(x_1 - x_2)^2 + (y_1 - y_2)^2}$.

In order to use this formula, take the two points (9, –2) and (3, 5) and treat them as (x_1, y_1) and (x_2, y_2), respectively.

Substitute the values into the formula in order to find the distance between the points. $d = \sqrt{(x_1 - x_2)^2 + (y_1 - y_2)^2}$

Evaluate within the parentheses: $d = \sqrt{(9 - 3)^2 + (-2 - 5)^2}$

Apply the exponent: $d = \sqrt{(6)^2 + (-7)^2}$

Add under the square root symbol: $d = \sqrt{36 + 49}$

$d = \sqrt{85}$

The distance between the points is $\sqrt{85} \approx 9.2$ units. If you chose answer choice (A), you may have subtracted the two values under the radical sign after they were squared. Choice (C) is the result of adding the values of x_1 and x_2, and the values of y_1 and y_2, instead of subtracting them as stated in the formula. Choice (D) is the result of subtracting the two values under the radical sign after they were squared and omitting the square root symbol. Choice (E) would be the result of omitting the square root symbol in the formula.

6. C

The distance between any two points in the coordinate plane can be found by using the distance formula.

In order to use this formula, first locate the points indicated in the graph. The endpoints of the line segment are located at R(1, 2) and S(–1, –2). Take the two points R(1, 2) and S(–1, –2) and treat them as R(x_1, y_1) and S (x_2, y_2), respectively.

Substitute the values into the formula in order to find the distance between the points. $d = \sqrt{(x_1 - x_2)^2 + (y_1 - y_2)^2}$

Evaluate within the parentheses: $d = \sqrt{[1-(-1)]^2 + [2-(-2)]^2}$

Apply the exponents: $d = \sqrt{(2)^2 + (4)^2}$

Add under the square root symbol: $d = \sqrt{4 + 16}$

$\sqrt{20} \approx 4.5$

The distance between the points is $\sqrt{20} \approx 4.5$ units. Choice (A) is the result of using the values of the first point for the x-values in the formula and the values of the second points for the y-values. If you chose (B), you may have subtracted the two values under the radical sign after they were squared.

Choice (E) would be the result of omitting the square root symbol in the formula.

7. C

In order to find the slope between any two points, substitute the values for each coordinate into the formula and simplify.

The first step is to take the formula for the slope of the line and substitute the values. Use the point (8, –4) as (x_1, y_1) and (–13, 10) as (x_2, y_2).

$$m = \frac{y_1 - y_2}{x_1 - x_2}$$

Substitute into the formula: $m = \dfrac{-4 - 10}{8 - (-13)}$

Simplify the numerator and denominator: $m = \dfrac{-14}{21}$

Simplify the fraction: $m = -\dfrac{2}{3}$

The slope of the line between the two points is $-\frac{2}{3}$. Choice (B) is the reciprocal of the correct choice. If you selected choice (D), you found a positive value and the slope of this line is negative. Choice (E) is the opposite of the reciprocal of the correct choice. Either of these may have been chosen if the *x*-values were placed in the numerator, instead of the denominator, and the *y*-values reversed as well.

8. D

In order to find the slope between any two points, substitute the values for each coordinate into the formula and simplify. In this case, locate two points on the line by looking at the graph. The *y*-intercept is at the point (0, 1) and another point on the line is (–1, 4).

The first step is to take the formula for the slope of the line and substitute the values of the two known points. Use the point (0, 1) as (x_1, y_1) and (–1, 4) as (x_2, y_2).

$$m = \frac{y_1 - y_2}{x_1 - x_2}$$

Substitute into the formula: $m = \dfrac{1 - 4}{0 - (-1)}$

Simplify the numerator and denominator: $m = \frac{-3}{1}$

Simplify the fraction: $m = -3$

The slope of the line between the two points is –3. Choice (A) is the reciprocal of the correct choice. Choice (B) is the opposite of the reciprocal of the correct choice. Choice (C) is the y-intercept of the line. If you selected choice (E), this value is positive and the slope in this case is negative.

9. D

The y-intercept of the graph of a line is the location where the line crosses the y-axis. In this figure the line crosses the y-axis at the point (0, –4). Thus, the y-intercept is –4. Choice (A) is the opposite of the slope of the line. Choice (B) may have been selected if the slope of the line was found, instead of the y-intercept. The correct slope of the line is $-\frac{1}{3}$. Choice (E) is the opposite of the y-intercept, and choice (C) is the reciprocal of this value.

10. B

First, change the given equation to slope–intercept form, or $y = mx + b$ form. In order to change the equation $2y - 2 = 6x$:

First, add 2 to each side of the equation: $2y - 2 + 2 = 6x + 2$

The equation simplifies to: $2y = 6x + 2$

Divide each side of the equation by 2: $\frac{2y}{2} = \frac{6x}{2} + \frac{2}{2}$

Simplify the equation: $y = 3x + 1$

In this equation, the value of $m = 3$ so the slope of the line is 3. The value of $b = 1$, so the y-intercept is 1. The correct answer choice is the graph of a line with a slope of 3 and a y-intercept of 1. Therefore, the correct answer is choice (B).

Choice (A) is the graph of the equation y = 2. Choice (C) is the graph of the equation $x = 2$. Choice (D) is the graph of the equation $y = 3x + 2$. Choice (E) is the graph of the equation $y = -3x + 1$.

CHAPTER 9

Graphing Equations on the Coordinate Plane

WHAT ARE GRAPHED EQUATIONS?

In the last chapter, you were introduced to the coordinate plane and the slope-intercept form of a linear equation. There are many other types of equations. You can identify the *graph*, or picture, of an equation according to its form. This chapter will explore some of these common equations and the graphs associated with them.

CONCEPTS TO HELP YOU

This section will describe four different types of equations and the corresponding picture that is created on the coordinate plane. The form that the equations are written in determines some properties of their graphs.

Point-Slope Form of a Linear Equation and Its Graph

In Chapter 8, the slope-intercept form of a linear equation was described. Another form that a linear equation can take is the *point-slope form.* This form enables you to quickly identify any point on the graph, represented as (x_1, y_1) and the corresponding slope, m.

THE POINT-SLOPE FORM OF A LINEAR EQUATION

The point-slope form of a linear equation is:

$y - y_1 = m(x - x_1)$

Where (x_1, y_1) is a point on the line, and m is the slope of the line.

Linear equations are always pictured as straight lines on the coordinate plane. Points on the plane, and the concept of slope, were reviewed in detail in Chapter 8. If you are given an equation in point-slope form, you can identify the slope and a point on the line. For example, if the equation is y –

$5 = 2(x - 3)$ the slope of the line is 2 and one point on the line is (3, 5). For the equation $y + 2 = -\frac{1}{2}(x + 1)$, the slope is $-\frac{1}{2}$ and a point on the line is (−1, −2). Take note that the coordinates of the point on the graph have the opposite sign to what is shown in the equation form.

Likewise, if you are given the graph of a line, you can find two points on the line, calculate the slope, and then create the equation using the slope and the coordinates of one of the points.

Absolute Value Linear Equations and Their Graphs

Absolute value linear equations are recognized by the presence of the absolute value symbols, for example $y = |x|$. This chapter will consider absolute value linear equations of the form $y = |mx + b|$. The absolute value of a number is always positive. Any equation of the form $y = |mx + b|$ will have all positive values for y. Therefore, the graph of these specific types of equations will all have y-coordinate values that are positive, and all of the graph will be above or on the x-axis.

When $y = |x|$, this means that $y = x$ when $x \geq 0$ and $y = -x$ when $x < 0$. The graph of this equation is shown below:

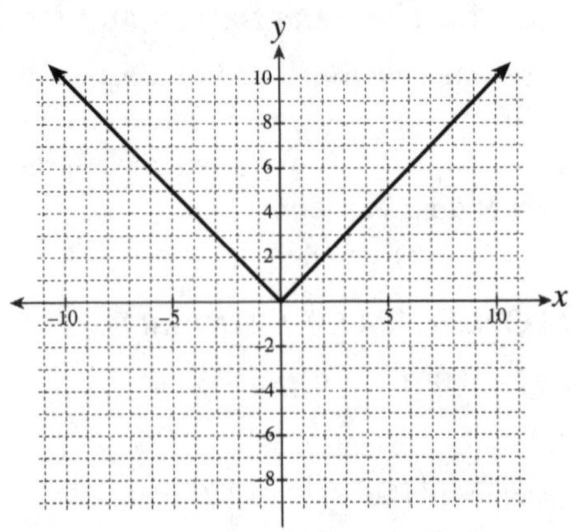

For absolute value linear equations, the basic form of the graph is a V shape, and for these equations of the form $y = |mx + b|$, all values are above or on the x-axis. Consider the equation $y = |2x + 3|$. This graph will be a combination of the graph of $y = 2x + 3$, and $y = -(2x + 3)$. The graph of this equation is shown below.

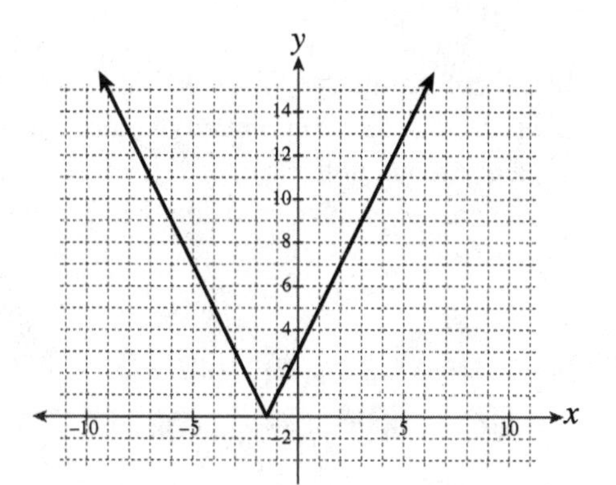

The point of the V will be at the *x-intercept*, where the graph crosses the x-axis, the value of x when $y = 0$. For this equation, the x coordinate can be found by these steps:

Write the equation, without the absolute value symbols: $0 = 2x + 3$

Subtract 3 from both sides of the equation: $0 - 3 = 2x + 3 - 3$

Simplify: $-3 = 2x$

Divide both sides by 2, and: $x = -\frac{3}{2}$

The point is the point of the V, the minimum point on the graph. You can verify this with the graph shown above.

QUADRATIC EQUATIONS AND THEIR GRAPHS

The standard form of a quadratic equation is:

$y = ax^2 + bx + c$

where a, b, and c are real numbers.

A quadratic equation is an equation of degree 2, meaning that its highest term is an x^2 term. The x^2 term of the equation is what makes it quadratic. When a quadratic equation is given in the form $y = ax^2 + bx + c$, the coefficients give important information about the graph.

Quadratic equation graphs are called *parabolas*, and they have a ∪ or ∩ a shape. If the leading coefficient, a, is positive, the shape is a ∪ and is said to open upward. If the leading coefficient, a, is negative, the shape is a ∩ and is said to open downward. These graphed equations have a line of symmetry, called the *axis of symmetry*, and a *vertex*, or *turning point*. The vertex is a point located on the axis of symmetry. When the leading coefficient, a, is positive, the vertex is also called the *minimum* of the graph. When a is negative, the vertex is a *maximum* of the graph. The constant term c is the y-intercept of the graph, where the graph crosses the y-axis. The vertex is on the axis of symmetry.

EQUATION FOR THE AXIS OF SYMMETRY OF A PARABOLA

The axis of symmetry has the linear equation: $x = \frac{-b}{2a}$

This is also the x-coordinate of the vertex.

A parabola for the equation $y = x^2 - 2x - 3$ is shown below:

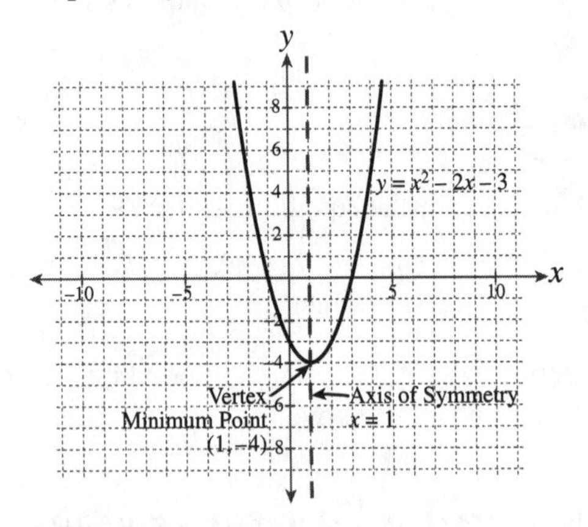

This graph has a "U" shape, and as you can see from the equation, the coefficient for the x^2 term is positive one. The axis of symmetry is $x = \frac{-b}{2a} = \frac{-(-2)}{2 \bullet 1} = \frac{2}{2}$, or $x = 1$. This is shown on the graph. The minimum point, the vertex, is (1, –4), also shown on the graph. This point can be found by substituting in the value of 1 for x to find the value of y: $y = 1^2 - 2(1) - 3 = 1 - 2 - 3 = -4$.

The Equation of a Circle and Its Graph

Circles were studied in depth in Chapter 4. This chapter will consider the graph of a circle on the coordinate plane.

THE EQUATION OF A CIRCLE

The standard form of an equation of a circle is:

$(x - h)^2 + (y - k)^2 = r^2$

Where (h, k) are the coordinates of the center of the circle, and r is the radius.

For example, the graph below shows two circles. Circle A has center at (–3, 2) with a radius of 2. This equation is $(x + 3)^2 + (y - 2)^2 = 2^2$, which simplifies to $(x + 3)^2 + (y - 2)^2 = 4$. Circle B has center at (6, 4) with a radius of 3. This equation is $(x - 6)^2 + (y - 4)^2 = 3^2$, which simplifies to $(x - 6)^2 + (y - 4)^2 = 9$.

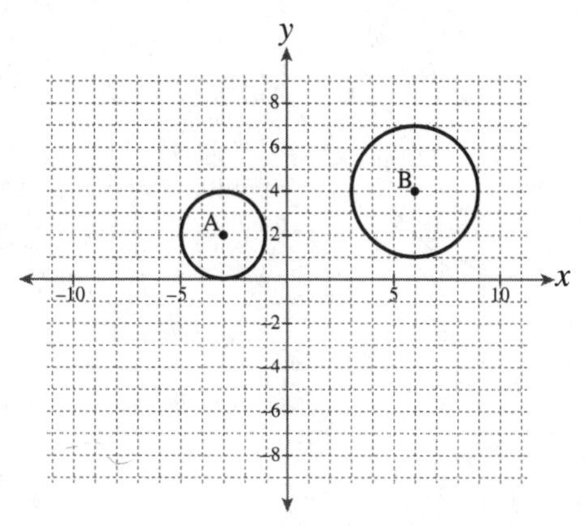

If you are given the center of a circle and any point on the circle, you can find the radius by using the distance formula, described in Chapter 8. The equation for the circle can now be determined. If you are given the coordinates of the endpoints of a diameter of a circle, you can calculate the center, which is the midpoint of the diameter. Midpoint is described in Chapter 8. Once you have the coordinates of the center, you can determine the radius and arrive at the equation of the circle.

Systems of Equations on the Coordinate Plane

A system of equations is two or more equations. To solve a system of equations means to find the x and y values that will make each equation true. You can solve a system of equations by using algebra, but this chapter will review how to solve a system of two equations graphically. Generally speaking, the solution to a system of equations is found in the coordinates of the point(s) where the graphs intersect.

There are three cases for a system of linear equations:

1. There is one solution, the point of intersection of the graphed lines.
2. There is no solution when the graphed lines are parallel and never intersect.
3. There is an infinite number of solutions when the graphed lines are equivalent.

These cases are illustrated below:

1.

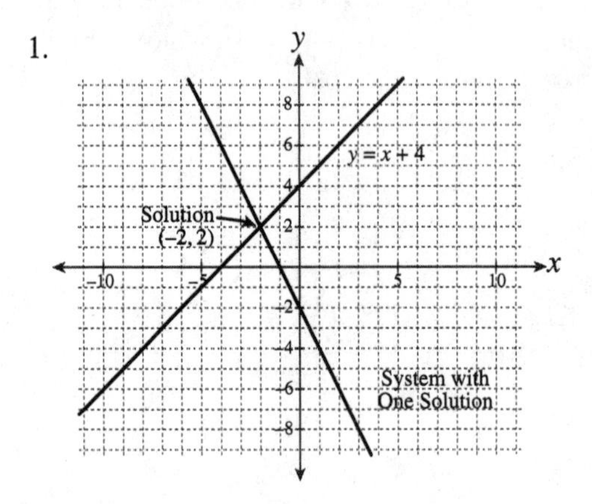

2.

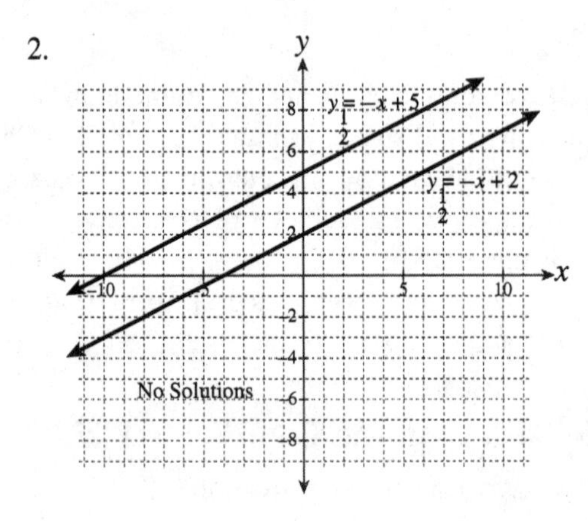

3.

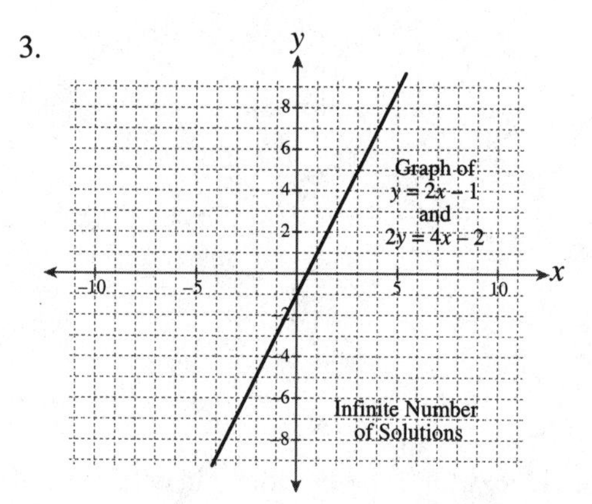

If a system consists of a linear equation and a quadratic equation, or a linear equation and a circle, there will be zero, one, or two solutions, as shown in the figure below:

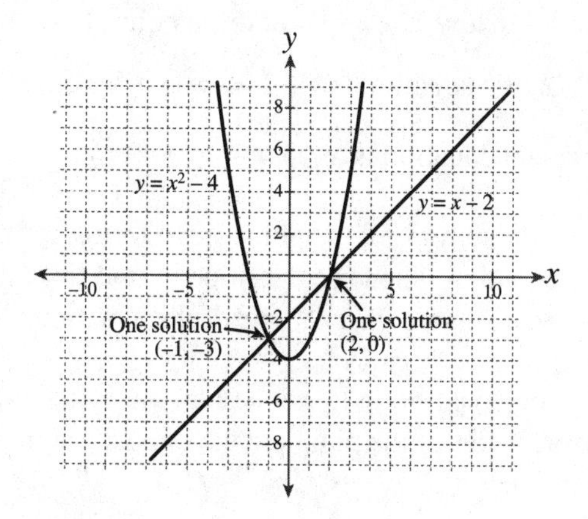

STEPS YOU NEED TO REMEMBER

This section will walk you through different types of problems concerning equations and their graphs. Study the examples carefully to master these concepts.

Identifying the Point-Slope Form of an Equation and Its Graph

If you are given the graph of a line on a coordinate plane, you can determine the point-slope form of its equation. First, identify the slope and one point on the graph. Consider the graph below:

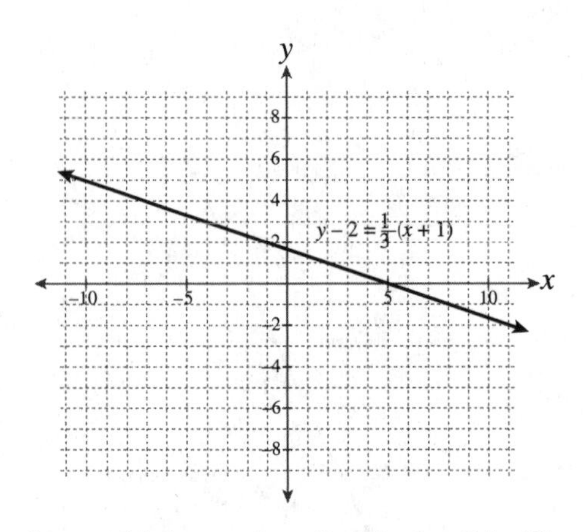

To calculate the slope, which was described in detail in Chapter 8, find two points on the graph and use the formula $m = \frac{y_1 - y_2}{x_1 - x_2}$. One point on the graph above is (−1, 2) and another point is (−4, 3). The slope is $\frac{3-2}{-4-(-1)} = \frac{1}{-3} = -\frac{1}{3}$. Now, use either of these points to write the point-slope form. Using (−1, 2) the equation is $y - 2 = -\frac{1}{3}(x+1)$. Using the other point, (−4, 3) the equivalent equation is $y - 3 = -\frac{1}{3}(x+4)$.

To graph a linear equation given the point-slope form, first determine the point used in the equation. Then use the slope to find two other points. Follow the steps below using the example of the equation $y + 3 = 2(x - 2)$:

1. Identify the given point. In this equation, the point is (2, −3). Notice that the coordinates of the point are the opposite sign to what is shown in the equation.
2. Graph this point.
3. Use the slope, in this case 2, to count up and over from this point. Since the slope is positive 2, count up two units and over to the right one unit to find another point. This will be point (3, −1). Plot this point.
4. Find a third point by starting at (3, −1) and counting up two units and over one unit to the right. This point is (4, 1). Plot this point.
5. Connect the points with a straight edge and label the line.

Here is the graph of $y + 3 = 2(x - 2)$:

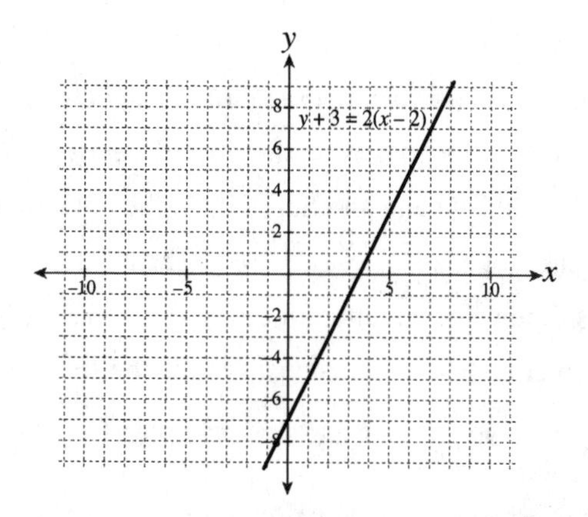

Identifying an Absolute Value Linear Equation and Its Graph

You can identify an absolute value linear equation of the form $y = |mx + b|$, by its V shape, where all points are on or above the x-axis. If you are given a graph of an absolute value equation, you can determine the equation by finding the equation of the right-hand portion of the graph. Look at the graph below:

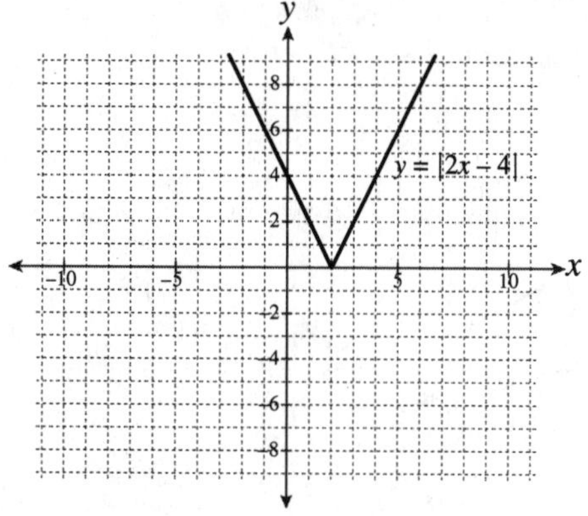

Extend the right-hand portion of the graph to find the y-intercept:

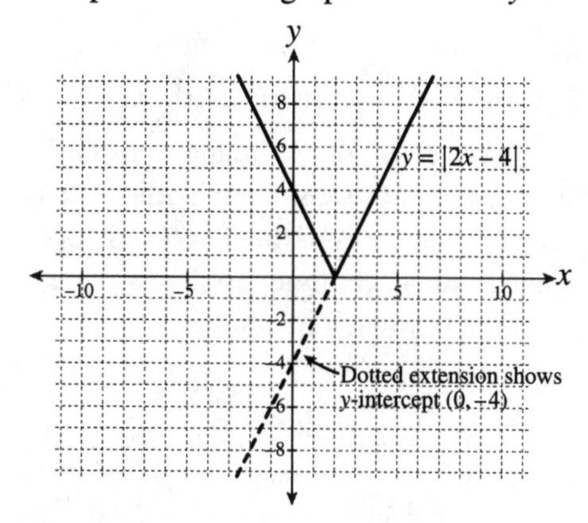

You can see from this extension that the y-intercept is –4. From there, determine the slope of this line, using the formula $m = \frac{y_1 - y_2}{x_1 - x_2}$, and two points. One point is (0, –4) and another point is the point of the V, at (2, 0). Substituting in, the slope is $\frac{-4-0}{0-2} = \frac{-4}{-2} = 2$. The equation is $y = |2x - 4|$.

To graph an absolute value equation of the type $y = |mx + b|$, remember that the graph will never be below the x-axis. Follow the steps below to graph an absolute value linear equation, using the example of: $y = \left|\frac{1}{2}x + 2\right|$

1. Find the x-intercept of the equation, the value of x when $y = 0$. In this example, the x-intercept is –4, because $0 = \frac{1}{2} \bullet -4 + 2$.
2. Graph the equation $y = \frac{1}{2}x + 2$, for all values when $x \geq -4$.
3. Graph the equation $y = -\left(\frac{1}{2}x + 2\right)$, which is $y = -\frac{1}{2}x - 2$, for all values when $x < -4$.

The graph of this equation is shown here:

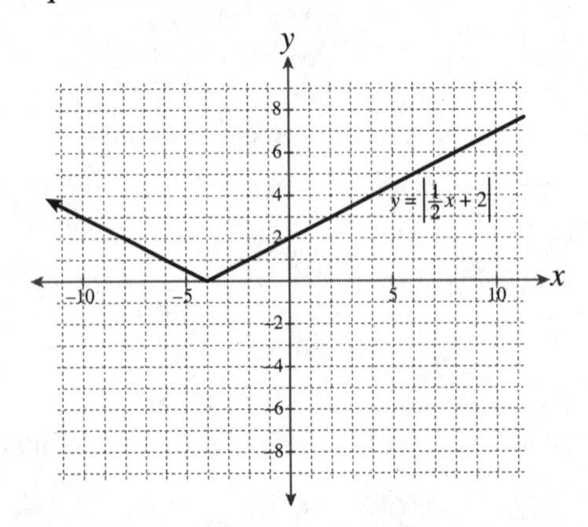

Identifying a Quadratic Equation and Its Graph

A quadratic equation has a graph that is called a *parabola*. You can identify a quadratic equation on a coordinate system by the ⋃ or ⋂ shape. From the graph, you can determine the equation of the axis of symmetry, the coordinates of the maximum or minimum point, and the y-intercept. Consider the graph of the following quadratic:

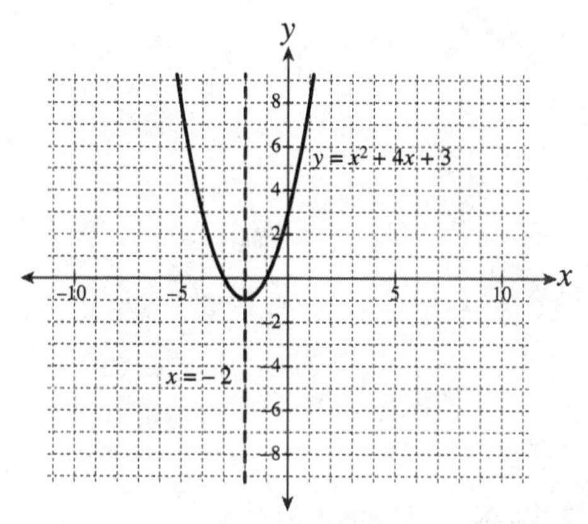

The graph crosses the y-axis at (0, 3), so the y-intercept is 3. This is also the value of c in the standard form of a quadratic $y = ax^2 + bx + c$. The axis of symmetry, shown above as a dotted line has the equation $x = -2$. From the figure, the minimum point is on this axis, and is $(-2,-1)$. You can determine the equation of a parabola, using algebra, but this is beyond the scope of this book.

If you are given a quadratic equation and asked to make a graph, the common method is to make a table of values, graph the coordinates, and make the smooth curve parabola. The intent of this chapter is not to review how to graph using a table method. But if you are given a quadratic equation and asked to identify its graph, use the axis of symmetry and the vertex to identify it. For example, find which graph below is the parabola for the quadratic equation $y = x^2 + x - 2$.

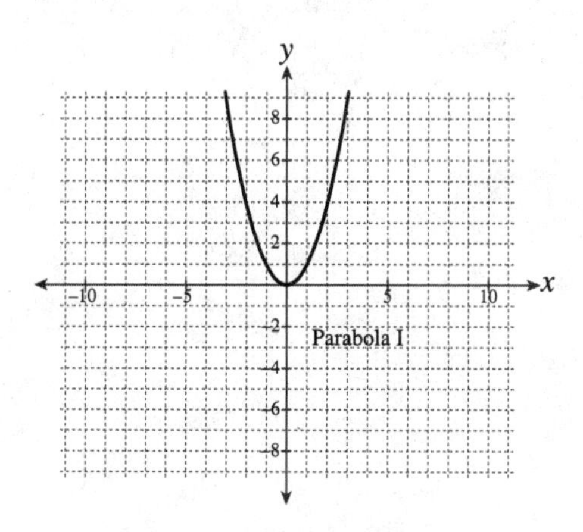

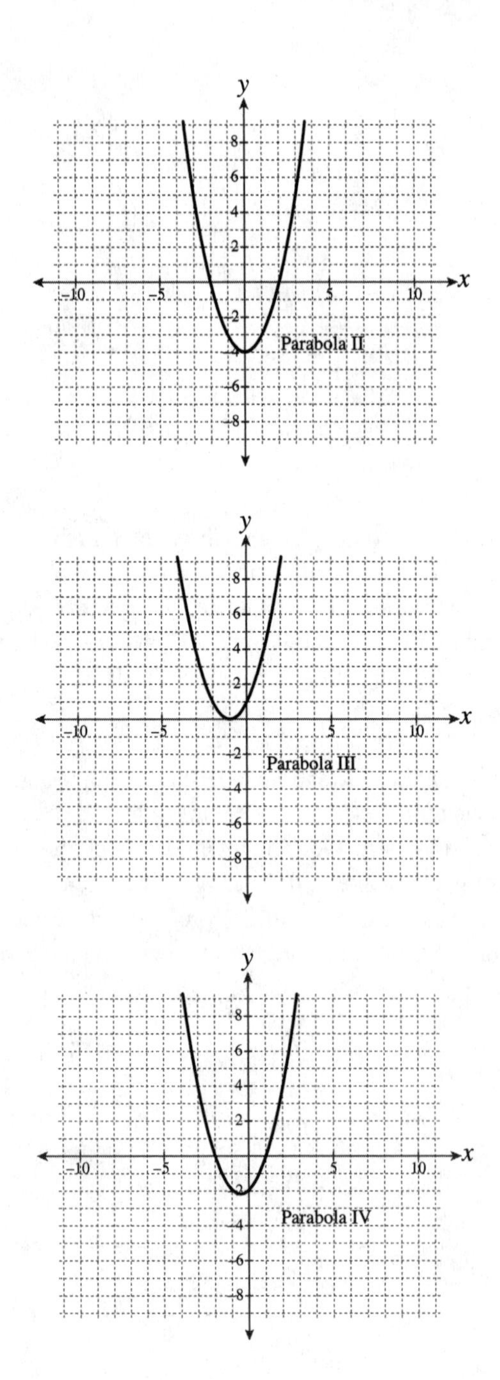

y
x
8
6
4
2
–10
–5
5
10
–2
–4
–6
–8
Parabola II
Parabola III
Parabola IV

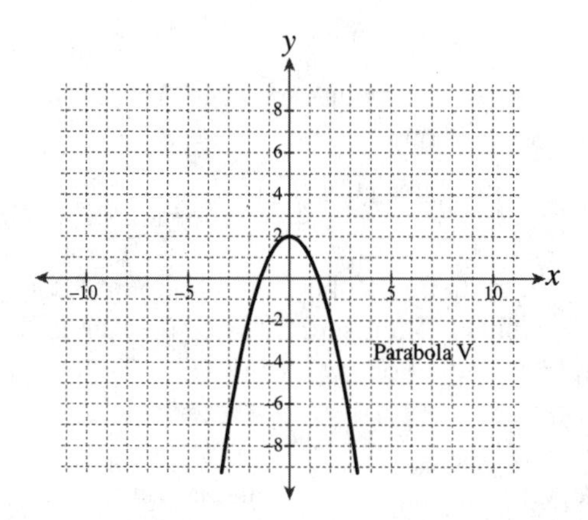

First, look at the choices. Graph V has a ∩ which would indicate a negative value for a in the standard form $y = ax^2 + bx + c$. The leading coefficient of this equation is positive, so choice V can be eliminated. Now, look for a graph with a y-intercept of –2, identified as the c value in the equation. This eliminates two of the graphs. Graph I has a y-intercept of 0, and graph III has a y-intercept of 1. Now, calculate the axis of symmetry for the equation. Use the formula $x = \frac{-b}{2a}$. Substitute in to get the equation $x = \frac{-1}{2 \cdot 1}$, or $x = -\frac{1}{2}$. The vertex will be at this x-coordinate. Graph II has an axis of symmetry at $x = 0$. Therefore, the graph of $y = x^2 + x - 2$ is graph IV.

Describing a Circle and Its Graph

If you are given a graph of a circle, you can determine its equation. First, identify the coordinates of the center and then determine the radius. Usually, you can count over from the center the number of spaces to get to a point on the circle. This number of spaces is the radius. Look at this graph of a circle with diameter AB:

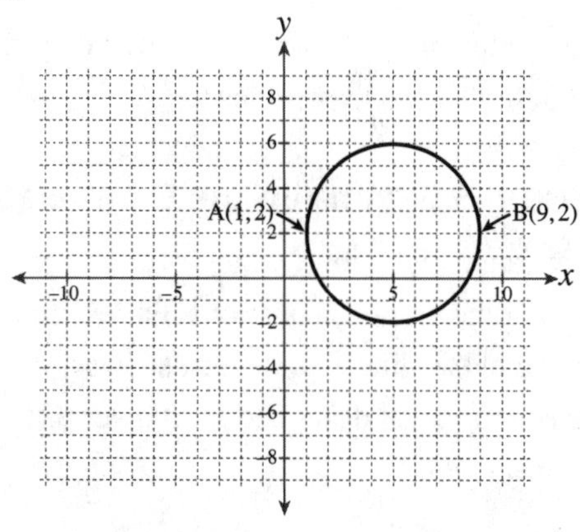

The center is halfway between points shown at A(1, 2) and B(9,2). So the center is at the midpoint of these points:

$\left(\frac{x_1+x_2}{2}, \frac{y_1+y_2}{2}\right) = \left(\frac{1+9}{2}, \frac{2+2}{2}\right) = (5, 2)$. The radius is the length of the segment from the center to one of these points. Count the spaces from the center, and the radius is 4 units long. Using the standard form of the equation of a circle, the equation is $(x-5)^2 + (y-2)^2 = 4^2$, or $(x-5)^2 + (y-2)^2 = 16$.

To graph a circle, given its equation in standard form, determine the center and the radius. Graph the center point. Plot four points emanating from this center counting the number of spaces equal to the radius going up, down, left, and right. Use a compass to sketch the circle, or estimate and sketch the curve of the circle using the four anchor points.

Another problem you may encounter is similar to the following:

> Find the equation of a circle with a center of (2, 3), passing through the point (4, 6).

You are told the center coordinates, but not the radius. The radius is the distance from the center to any point on the circle. Use the distance formula to calculate the radius. Let (x_1, y_1) be (2, 3) and (x_2, y_2) be (4, 6).

Write the formula: $r = \sqrt{(x_1 - x_2)^2 + (y_1 - y_2)^2}$

Substitute in the values: $r = \sqrt{(2-4)^2 + (3-6)^2}$

Evaluate within parentheses: $r = \sqrt{(-2)^2 + (-3)^2}$

Evaluate the exponents and add: $r = \sqrt{13}$

Now you know the center and the radius. The equation is therefore $(x-2)^2 + (y-3)^2 = (\sqrt{13})^2$, which simplifies to $(x-2)^2 + (y-3)^2 = 13$.

Solving Systems of Equations on the Coordinate Plane

The solution to a system of equations, when solved graphically, is the intersection point(s) of the graphed equations. For example, to find the solution to the system of two linear equations, such as $y = \frac{1}{2}x + 1$ and $y = -3x - 6$, graph the equations, as described in Chapter 8. The graphs are shown here:

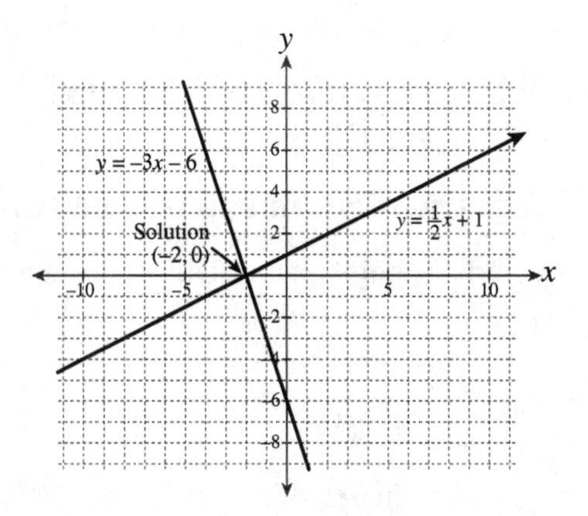

Look at the graph; the solution is where the graphed lines meet, at (−2, 0). This is the point that will make each of the equations true.

To solve a system of a circle and a line, name the points of intersection. Remember that there may be zero, one, or two points of intersection. The graph below shows a circle whose equation is $x2 + y2 = 25$, and a line whose equation is $y = -x + 5$. The solution is the two points of intersection, (0, 5) and (5, 0).

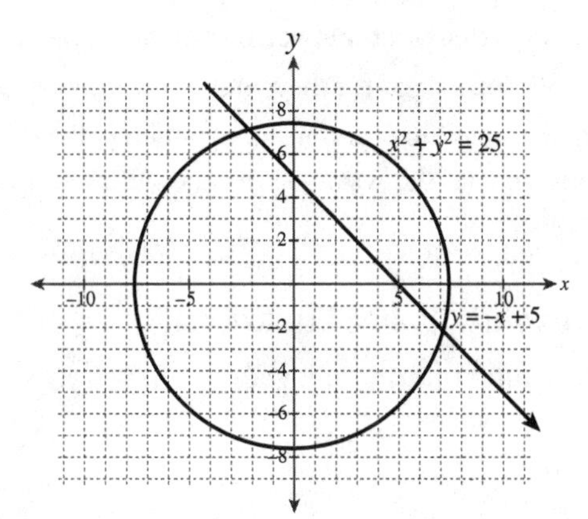

STEP-BY-STEP ILLUSTRATION OF THE FIVE MOST COMMON QUESTION TYPES

Question 1: Linear Equations in Point-Slope Form

Name the slope and one point on the graph of the equation
$y - 4 = \frac{1}{2}(x + 2)$

(A) Slope = $\frac{1}{2}$, and a point (2, –4)

(B) Slope = $\frac{1}{2}$, and a point (–2, 4)

(C) Slope = –4, and a point $\left(\frac{1}{2}, 2\right)$

(D) Slope = 2, and a point (–2, 4)

(E) Slope = 2, and a point (4,–2)

The point-slope form of a linear equation is $y - y_1 = m(x - x_1)$, where m is the slope of the line, and (x_1, y_1) is a point on the graph. For this equation, $m = \frac{1}{2}$, $x_1 = -2$ and $y_1 = 4$, so **the correct answer is choice (B).** If you chose choice (A), you used the incorrect signs for the x and y coordinates of the point. Choice (D) has the wrong value for the slope, which is the reciprocal of the actual slope. Choice (E) has the wrong slope and the x and y coordinates of the point are switched.

Question 2: Absolute Value Equations

What is the equation of the graph shown?

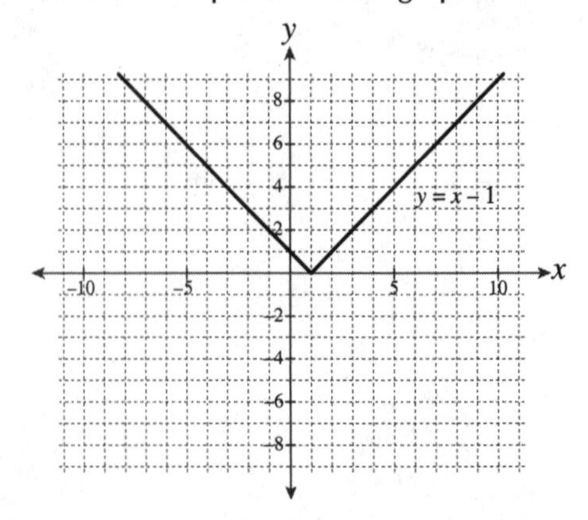

(A) $y = x - 1$

(B) $y = -x + 1$

(C) $y = |x - 1|$

(D) $y = |x + 1|$

(E) $y = |x| - 1$

The graph shows a V shape, so it is an absolute value linear equation. Extend the right-hand side of the V to find the y-intercept of this piece of the graph.

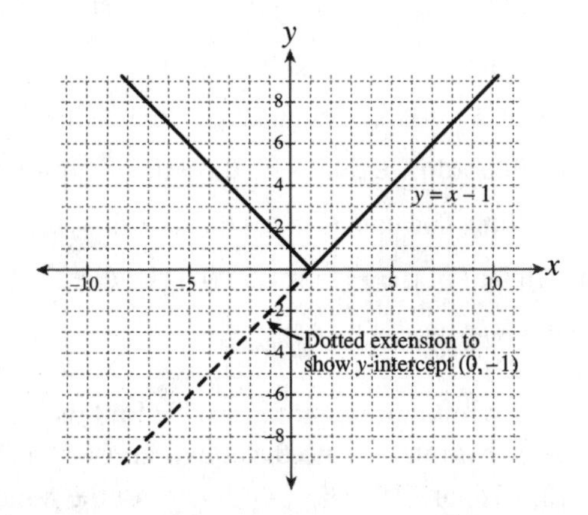

As shown above, the y-intercept is –1. Find two points on the right-hand side of the graph, such as $(x_1, y_1) = (1, 0)$ and $(x_2, y_2) = (2, 1)$. Calculate the slope of this portion: $m = \frac{y_1 - y_2}{x_1 - x_2} = \frac{0 - 1}{1 - 2} = \frac{-1}{-1} = 1$. The graph of this equation is on or above the x-axis, so it is an absolute value linear equation of the form $y = |mx + b|$, where m is the slope of the line and b is the y-intercept. **The correct answer choice is (C).**

Answer choice (A) is the equation of the graph of a line with slope of 1 and y-intercept of –1, not an absolute value equation. Choice (B) is not an absolute value equation either; it is the equation of a line with slope of –1 and y-intercept of 1. Answer choice (D) has the incorrect y-intercept value for the right-hand portion. Answer choice (E) is the graph of $y = |x|$, shifted down one unit; this equation would have one solution with a y value that is less than 0, which is not the graph shown.

Question 3: Finding the Minimum Value for a Quadratic Equation

Identify the minimum value for the equation $y = x^2 + 2x - 15$.

(A) (0,–15)

(B) (–1,–16)

(C) (–1,–12)

(D) (1,–12)

(E) (–5, 0)

The minimum value of a quadratic equation is also called the *vertex.* The vertex point has an x-coordinate that is on the axis of symmetry for the parabola of the equation. The equation for the axis of symmetry is $x = \frac{-b}{2a}$, where a is the coefficient of the x^2 term and b is the coefficient of the x term. Substitute in to get $x = \frac{-2}{2 \cdot 1} = \frac{-2}{2} = -1$

The x-coordinate of the minimum point is –1. To find the y-coordinate of the minimum point, substitute –1 into the equation: $y = (-1)2 + 2(-1) - 15$, which is 1 + –2 – 15, or –16. The coordinates of the minimum value is (–1,–16), so **the correct answer choice is (B).** Choice (A) is the y-intercept of the parabola. If you chose answer (C), you calculated the x-coordinate correctly, but then you made an addition error when calculating the corresponding y-coordinate. If your choice was (D), you forgot the negative sign in the formula for the equation of the axis of symmetry. Choice (E) is one of the x-intercepts of the parabola—a value for x when $y = 0$.

Question 4: Finding the Equation of a Circle

What is the equation of a circle with center at (–2, 3) and with one point on the circle at (2, 3)?

(A) $(x + 2)^2 + (y - 3)^2 = 16$

(B) $(x - 2)^2 + (y + 3)^2 = 16$

(C) $(x + 2)^2 + (y - 3)^2 = 4$

(D) $(x - 2)^2 + (y + 3)^2 = 4$

(E) $(x + 2)^2 + (y - 3)^2 = 64$

The equation of a circle is $(x - h)^2 + (y - k)^2 = r^2$, where (h, k) is the center of the circle, and r is the radius length. The radius length is the distance from the center of the circle to any point on the circle. Use the distance formula to calculate the radius. The formula is $r = \sqrt{(x_1 - x_2)^2 + (y_1 - y_2)^2}$, where (x_1, y_1) is the center, (–2, 3) and (x_2, y_2) is the point (2, 3).

Write the formula: $r = \sqrt{(x_1 - x_2)^2 + (y_1 - y_2)^2}$

Substitute in the values: $r = \sqrt{(-2 - 2)^2 + (3 - 3)^2}$

Evaluate within parentheses: $r = \sqrt{(-4)^2 + (0)^2}$

Evaluate the exponents and simplify: $r = \sqrt{16} = 4$

The radius length is 4. So the equation of this circle is $(x + 2)^2 + (y - 3)^2 = 4^2$, which simplifies to $(x + 2)^2 + (y - 3)^2 = 16$. **The correct answer choice is (A).** If your answer was (B), you used an incorrect form of the equation as $(x + h)^2 + (y + k)^2 = r^2$. If you chose (C), you forgot to square the radius. Choice (D) is the incorrect form of the equation and a result of not using the radius squared. Choice (E) would be the result of incorrectly giving the radius to the third power.

Question 5: Finding the Solution to a System of Equations

What is the solution to the system of graphed equations below?

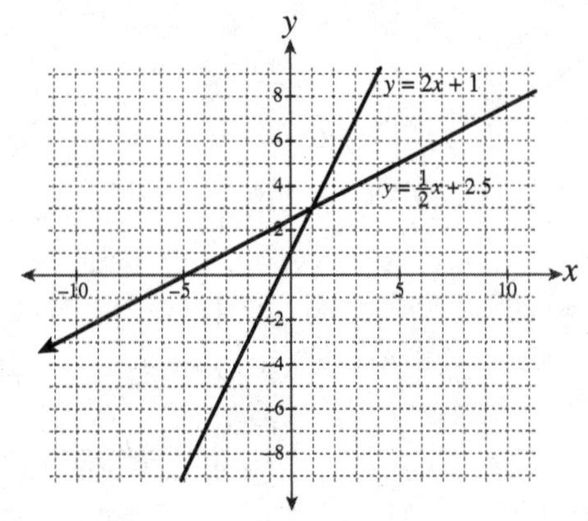

(A) (1, –3)

(B) (3, 1)

(C) (1, 3)

(D) (–3, –1)

(E) (–1, –3)

You can easily find the solution to a set of graphed linear equations by finding the coordinates of the point of intersection. In the figure above, the solution is (1, 3), the point where the two lines meet. **Choice (C) is the correct answer.** If your answer was (A), you found the incorrect *y*-coordinate for the solution. Choice (B) may have been incorrectly chosen if you switched the *x* and *y* coordinates. Choice (D) has both the coordinate values switched and they are also incorrectly shown as negative values. Choice (E) has the coordinate values as negative.

CHAPTER QUIZ

1. Which of the following points is on the graph of the equation $y - 3 = 4(x + 5)$?

   (A) (–5, 3)

   (B) (5, –3)

   (C) (3, –5)

   (D) (4, 5)

   (E) (–3, 5)

2. Given the graph of the linear equation below, what is the point-slope form of the equation?

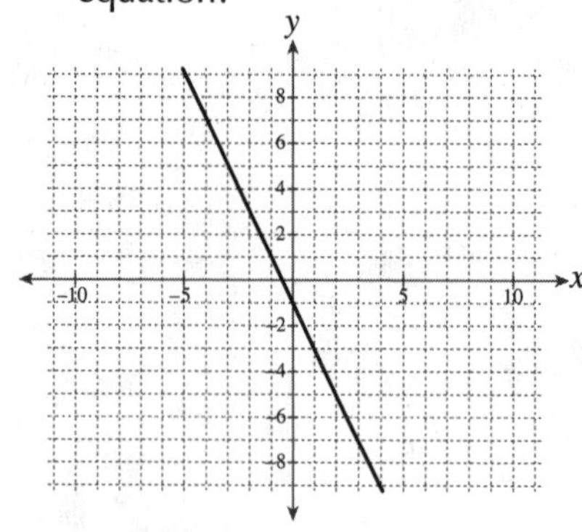

   (A) $y + 3 = 2(x - 1)$

   (B) $y - 3 = 2(x + 1)$

   (C) $y - 3 = -2(x + 1)$

   (D) $y + 3 = -2(x - 1)$

   (E) $y + 3 = \frac{1}{2}(x - 1)$

3. What is the equation of the following graph?

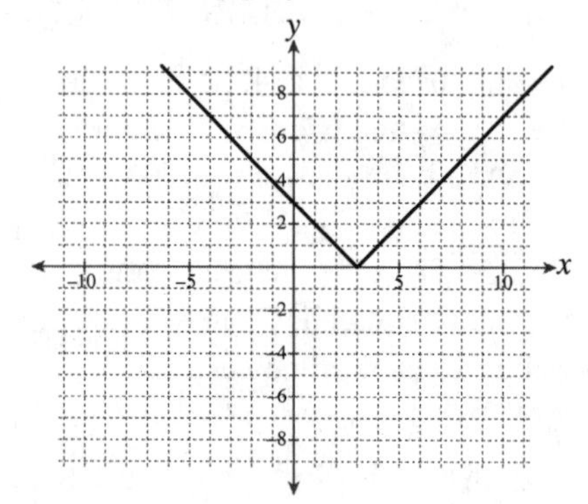

   (A) $y = |x|$

   (B) $y = |x - 3|$

   (C) $y = x - 3$

   (D) $y = -x + 3$

   (E) $y = x^2 + 3$

4. Which graph represents the equation $y = |2x - 2|$?

   (A)

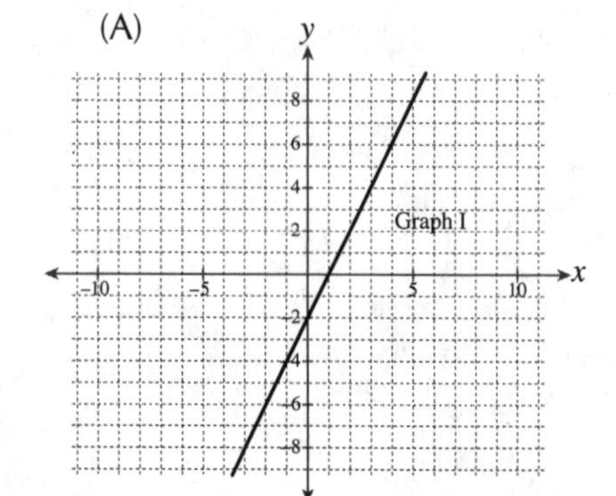

(B)

(E)

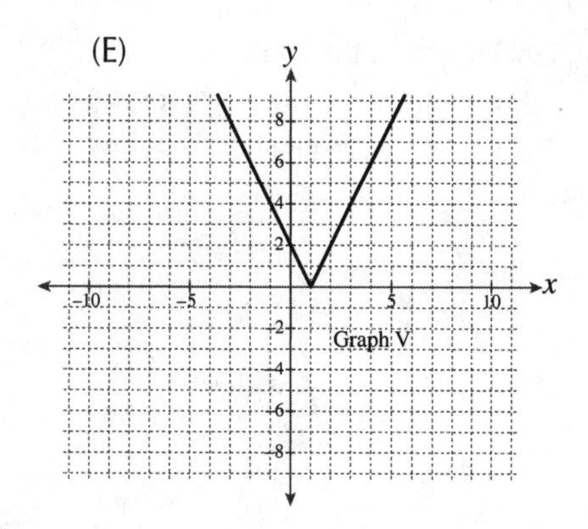

(C)

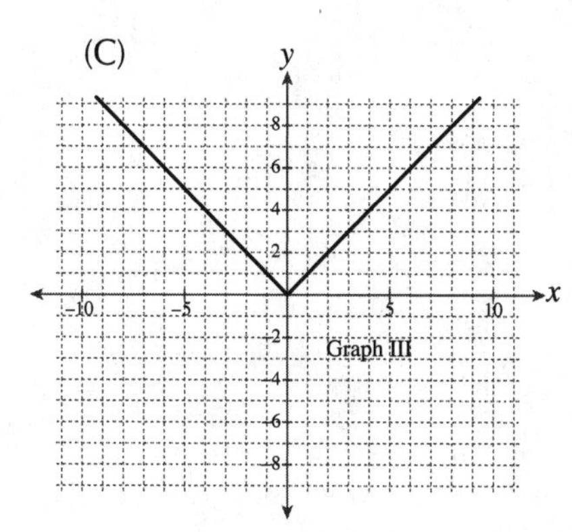

(D)

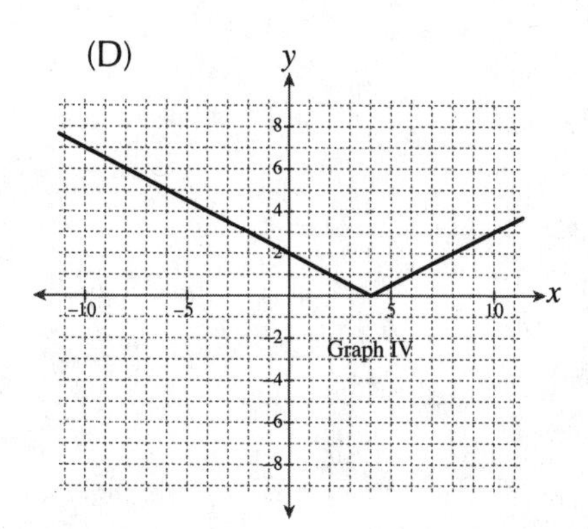

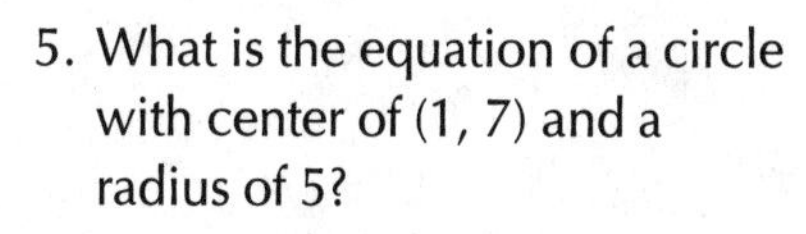

5. What is the equation of a circle with center of (1, 7) and a radius of 5?

(A) $(x + 1)^2 + (y + 7)^2 = 5$

(B) $(x - 1)^2 + (y - 7)^2 = 5$

(C) $(x - 1)^2 + (y - 7)^2 = 25$

(D) $(x + 1)^2 + (y - 7)^2 = 25$

(E) $(x - 1)^2 + (y - 7)^2 = 10$

6. What is the equation of a circle with a diameter that has endpoints at (–1, 2) and (5, 2)?

(A) $(x - 3)^2 + (y - 2)^2 = 9$

(B) $(x - 2)^2 + (y - 2)^2 = 9$

(C) $(x + 2)^2 + (y + 2)^2 = 9$

(D) $(x - 2)^2 + (y - 2)^2 = 3$

(E) $(x + 2)^2 + (y + 2)^2 = 3$

7. What is the vertex point of the graph of the equation $y = x^2 - 4x - 5$?

(A) (2, –9)

(B) (–2, 7)

(C) (0, –5)

(D) (5, 0)

(E) (–1, 0)

8. Which of the following equations could be the equation of the parabola shown below?

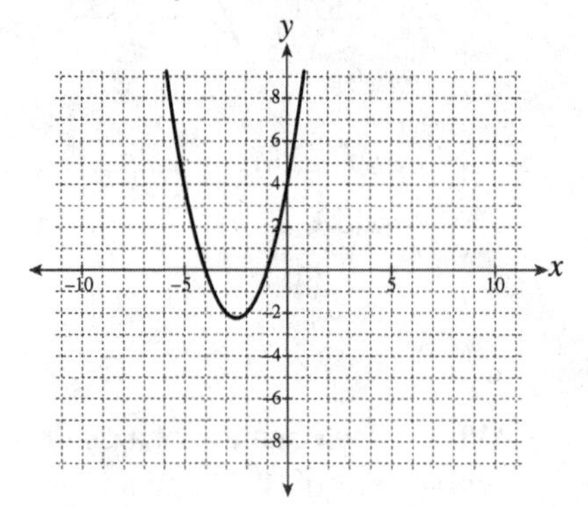

(A) $y = x^2 + 2.5x + 4$

(B) $y = -x^2 + 5x + 4$

(C) $y = -x^2 - 5x - 8.5$

(D) $y = x^2 - 5x + 4$

(E) $y = x^2 + 5x + 4$

9. What is the solution to the following system of graphed equations?

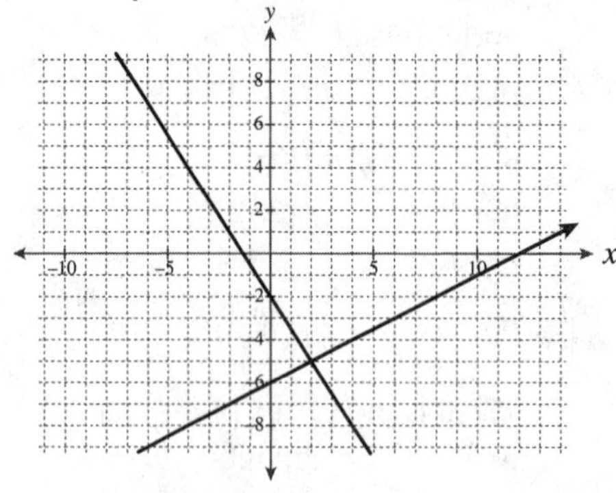

(A) (–2, –5)

(B) (–5, 2)

(C) (2, –5)

(D) (–5, –2)

(E) (2, –4)

10. What is a solution to the following system of graphed equations?

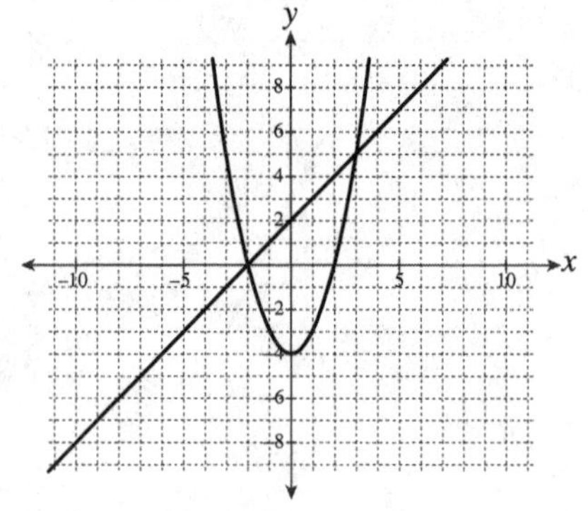

(A) (0, –2)

(B) (3, 5)

(C) (5, 3)

(D) (0, 2)

(E) (0, –4)

ANSWER EXPLANATIONS

1. A

When a linear equation is in point-slope form, you can quickly identify one of the points of the graph. The point-slope form is $y - y_1 = m(x - x_1)$, where m is the slope of the line, and (x_1, y_1) is a point on the graph. In this equation, $x_1 = -5$ and $y_1 = 3$. The point is (–5, 3), answer choice (A). If you chose answer (B), you used the wrong signs for the coordinates, a common error when interpreting these types of questions. If you chose (C), you switched the x and y coordinates. In choice (D), the value of the slope is the x-coordinate, which is incorrect. If you chose (E), you switched the x and y-coordinates and also used the wrong signs.

2. D

To determine the point-slope form of a linear equation, first find the slope of the graphed line, by using two points on the graph. You can use any two points on the graph, such as (0, –1) and (1, –3). The formula for the slope of a line, given two points, is $m = \frac{y_1 - y_2}{x_1 - x_2}$. Substitute into this formula to get $m = \frac{-1-(-3)}{0-1} = \frac{2}{-1} = -2$

The point-slope form of a linear equation is $y - y_1 = m(x - x_1)$. Using the point (1, –3) and the slope of –2, one form of the equation is $y + 3 = -2(x - 1)$. In choice (A), you incorrectly found the slope to be 2 instead of –2. In choice (B), you had the wrong signs for the slope and the x and y coordinates used to produce that equation. If your answer was choice (C), you used the incorrect signs of the point to produce the equation. In answer choice (E), you have calculated the slope incorrectly, by placing the change in x values in the numerator instead of the denominator.

3. B

This graph has the familiar V shape, which indicates an absolute value linear equation of the form $y = |mx + b|$. Take the right-hand side of the equation and determine the slope and the y-intercept. The slope of this portion can be determined by using two points, such as (3, 0) and (4, 1) and substituting into the formula $m = \frac{y_1 - y_2}{x_1 - x_2}$ to get $m = \frac{0-1}{3-4} = \frac{-1}{-1} = 1$. The

y-intercept can be determined by extending the right-hand portion down to the y-axis. This portion will cross the y-axis at –3, so the y-intercept is (0, –3), or $b = -3$. Now that you have established the slope and the y-intercept, the equation is $y = |x - 3|$. Answer choice (A) would be a graph where the y-intercept is 0. Choice (C) is the correct slope and y-intercept, but the absolute value symbols are not present. In choice (D), you have calculated the slope incorrectly, perhaps through a mistake in integer arithmetic. Choice (E) is a quadratic equation, which would have a "U" shape.

4. E

This equation is an absolute value linear equation. The graphs of these types of equations have a V shape. This equation is in the form $y = |mx + b|$, where m represents the slope of the right-hand portion of the line and b represents the y-intercept of the right-hand portion. Pick two points on the right-hand portion of the graph to determine the slope. Using points (1, 0) and (2, 2), calculate the slope. Use the formula $m = \frac{y_1 - y_2}{x_1 - x_2}$, and substitute in to get the formula $m = \frac{0-2}{1-2} = \frac{-2}{-1} = 2$. If you extend the right-hand portion of the graph down to the y-axis, you get the y-intercept of –2. The equation is therefore $y = |2x - 2|$. Choice (A) is the graph of the linear equation $y = 2x - 2$. Although it has the correct slope and y-intercept, it is not an absolute value graph. Choice (B) is the graph of the linear equation $y = -2x - 2$, and is not an absolute value graph. Choice (C) is the graph of $y = |x|$. The y-intercept of the right-hand portion is 0, not –2, and it also has a slope of 1. Choice (D) is the graph of the absolute value linear equation $y = \left|\frac{1}{2}x - 2\right|$. If this was your choice, you calculated an incorrect slope.

5. C

The equation of a circle takes the form $(x - h)^2 + (y - k)^2 = r^2$, where (h, k) is the center of the circle, and r is the radius length. The center, (h, k), is (1, 7), so $h = 1$ and $k = 7$. Given that the radius is 5, the correct equation is $(x - 1)^2 + (y - 7)^2 = 5^2$, which simplifies to $(x - 1)^2 + (y - 7)^2 = 25$. Answer choice (A) is a circle with center at (–1, –7) and radius of $\sqrt{5}$. If you chose choice (B), you forgot to square the radius in the equation. Choice (D) is a circle with radius of 5 but with a center at (–1, 7). If your choice was (E), you multiplied the radius by two, instead of squaring the radius.

6. B

The equation of a circle takes the form $(x - h)^2 + (y - k)^2 = r^2$, where (h, k) is the center of the circle, and r is the radius length. You need to determine the center coordinates and the length of the radius. You are given the endpoints of a diameter of the circle. The center will be the midpoint between these points. Use the midpoint formula and the given coordinates to get $\left(\frac{x_1 + x_2}{2}, \frac{y_1 + y_2}{2}\right) = \left(\frac{-1+5}{2}, \frac{2+2}{2}\right) = (2, 2)$ as the center of the circle. The radius is the length from the center to a point on the circle. Use the center coordinates and one of the points, such as (5,2) to substitute into the distance formula.

Write the formula: $r = \sqrt{(x_1 - x_2)^2 + (y_1 - y_2)^2}$

Substitute in the values: $r = \sqrt{(5-2)^2 + (2-2)^2}$

Evaluate within parentheses: $r = \sqrt{3^2 + 0}$

Evaluate the exponents and simplify: $r = \sqrt{9} = 3$

The radius is 3. The equation is therefore $(x - 2)^2 + (y - 2)^2 = 3^2$, which simplifies to $(x - 2)^2 + (y - 2)^2 = 9$.

7. A

The vertex point of a quadratic equation has its x-coordinate on the axis of symmetry. The equation for the axis of symmetry is $x = \frac{-b}{2a}$. In this equation, a is the coefficient associated with the x^2 term, and b is the coefficient associated with the x term. Substitute in the values to get the equation for the axis of symmetry which is $x = \frac{-(-4)}{2 \cdot 1} = \frac{4}{2} = 2$. The y-coordinate can be found by substituting in $x = 2$ into the original equation, $y = x^2 - 4x - 5$, or $y = 2^2 - 4(2) - 5 = 4 - 8 - 5 = -9$. The coordinates of the vertex are (2, –9).

If your answer was (B), you forgot the negative sign for the axis of symmetry formula. Answer choice (C) is the y-intercept of the equation. Answer choices (D) and (E) are each one of the solutions to the equation $0 = x^2 - 4x - 5$.

8. E

This graph is a parabola with a "U" shape. This indicates that the equation is a quadratic equation of the form $y = ax^2 + bx + c$. The coefficient a will be positive because of the shape. This eliminates answer choices (B) and (C). The coefficient c is the y-intercept of the parabola, in this case 4. Because all remaining equations have the value of $c = 4$, you need to determine the equation of the axis of symmetry to find the correct equation. The axis of symmetry is the vertical line that goes through the vertex, at the x-value of -2.5. Use the formula for the axis of symmetry, $x = \frac{-b}{2a}$, and try the remaining choices to see which has the axis at $x = -2.5$. For choice (A), the axis of symmetry is $x = \frac{-2.5}{2 \bullet 1} = -1.25$. For choice (D) the axis of symmetry is $x = \frac{-(-5)}{2 \bullet 1} = 2.5$. Choice (E), the correct equation has the axis of symmetry

at $x = \frac{-5}{2 \bullet 1} = -2.5$.

9. C

The solution to the system of equations is the coordinates of the point of intersection of the two graphs. This graphic shows two linear equations, and they meet at (2, –5), the solution to the system. Choice (A) has an incorrect negative sign on the x-coordinate. If you chose (B), you switched the x and y coordinates. Choice (D) has two errors—the x and y-coordinates are switched, and the 2 is –2. If your choice was (E) you probably counted down incorrectly to get a wrong value for the y-coordinate.

10. B

The solution to a system of equations that are graphed is found by identifying the point or points of intersection. This is a graph of a parabola and a line. There are two points of intersection, therefore two solutions. These points are (3, 5) and (–2, 0). Choice (A) is one of the solutions with the x- and y-coordinates switched. Choice (C) is the other solution with the x- and y-coordinates switched. Choice (D) is the y-intercept of the linear equation. Choice (E) is the y-intercept of the parabola.

CHAPTER 10

Transformational Geometry

WHAT IS TRANSFORMATIONAL GEOMETRY?

Transformational geometry is the study of changes on geometric figures. The type of changes that occur may flip the object, turn the object, slide the object, or shrink/enlarge the object. The terms associated with these changes are *reflections, rotations, translations,* and *dilations*. Each of these transformations will be explained in this chapter and examples given to help you understand the various situations you may encounter in your study.

CONCEPTS TO HELP YOU

When studying transformations, four major types emerge. Each of these types, along with concepts associated with them, will be introduced and illustrated in this section.

The Four Basic Types of Transformations

Line Reflections and Line Symmetry

One of the most common types of transformations is a line reflection. A *line reflection* occurs when a point or figure is reflected, or flipped, over a line. The reflected image is the same distance away from the line of reflection as the original figure but is located on the opposite side. The resulting image for any transformation is written using a "prime" symbol. For instance, the image of point P′ after a transformation is, which is read "P prime." In the figure below, triangle *ABC* is reflected over line *m* to become triangle *A'B'C'*.

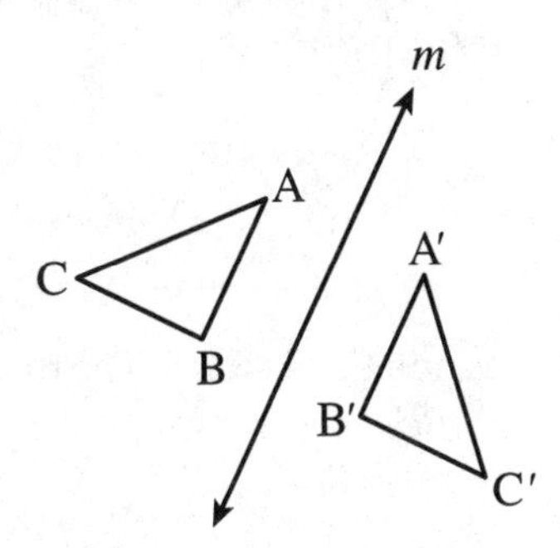

The symbol used to denote a reflection is R. The reflection of triangle ABC over line *m* is written as $\Delta ABC \xrightarrow{R_m} \Delta A'B'C'$.

Line symmetry occurs when a line can be drawn through a figure and a mirror image of each side appears on either side of the line. Depending on the figure, there can be one line of symmetry, many lines of symmetry, or no lines of symmetry. For example, take a look at the figures below. An isosceles triangle has one line of symmetry, a square has four lines of symmetry, and a parallelogram has no lines of symmetry.

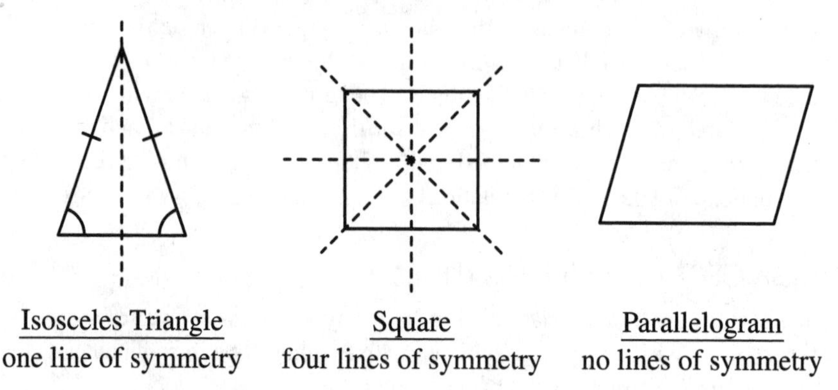

Isosceles Triangle — one line of symmetry
Square — four lines of symmetry
Parallelogram — no lines of symmetry

Rotations and Rotational Symmetry

A *rotation* of a geometric figure occurs when the figure is turned about a point in the plane. When the figure is rotated, the shape and size of the figure remains the same. The example below shows triangle *ABC* being rotated 180 degrees to become triangle *A'B'C'*. Notice that the shape and size are the same, but the triangle now appears upside down.

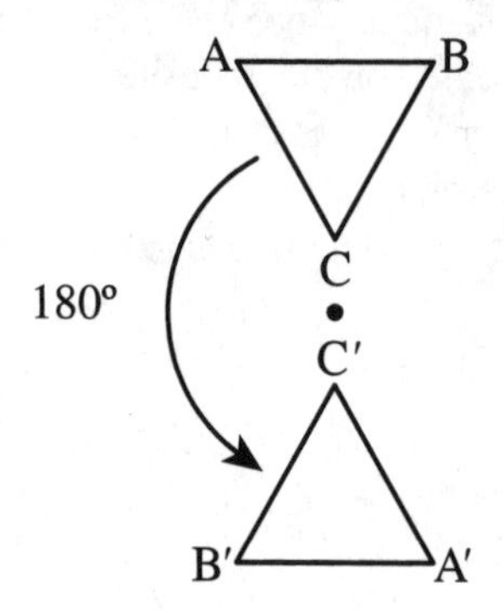

The symbol used to denote a rotation is ROT. The rotation of triangle ABC 180 degrees is written as $\Delta ABC \xrightarrow{T_{180}} \Delta A'B'C'$.

Rotational symmetry occurs when a figure can be rotated, or turned, a certain number of degrees less than 360 and appears to look the same as the original figure. Recall that a circle has 360 degrees; so turning a figure completely around would be turning the figure 360 degrees. The number of degrees of rotational symmetry is based on the number of times you see an image that is the same as the original figure. For example, take a regular hexagon. A regular hexagon has six congruent sides and six congruent angles. As shown in the figure below, if this polygon is rotated, you will see an image that appears the same as the original figure at six different times. Therefore, a regular hexagon has 60 degree rotational symmetry, since 360 $\div 6 = 60$.

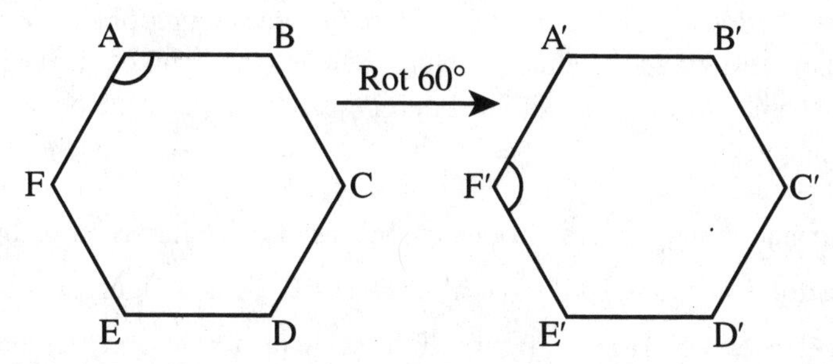

An equilateral triangle has three congruent sides and three congruent angles. Thus, an equilateral triangle has rotational symmetry of 120 degrees, since $360 \div 3 = 120$.

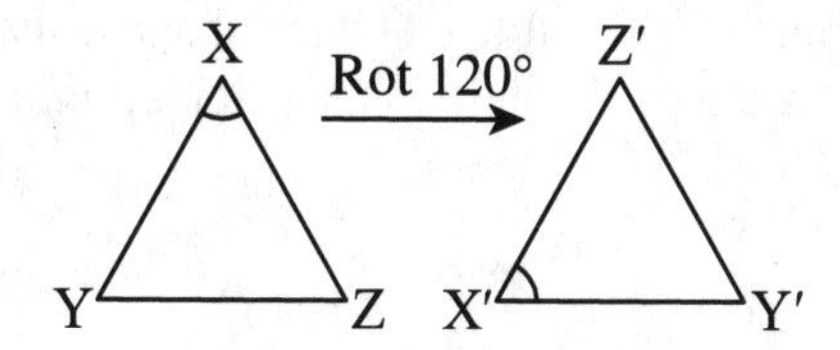

Translations

A *translation* occurs when each part of the figure being translated is moved, or shifted, to another location without turning or flipping the figure. The shape and size of the figure remain the same after it has been translated. A translation is also known as a *slide*. The example below shows triangle ABC being translated four units to the right and two units down to become triangle $A'B'C'$. Notice that the shape and size are the same, but the triangle now appears in a different location.

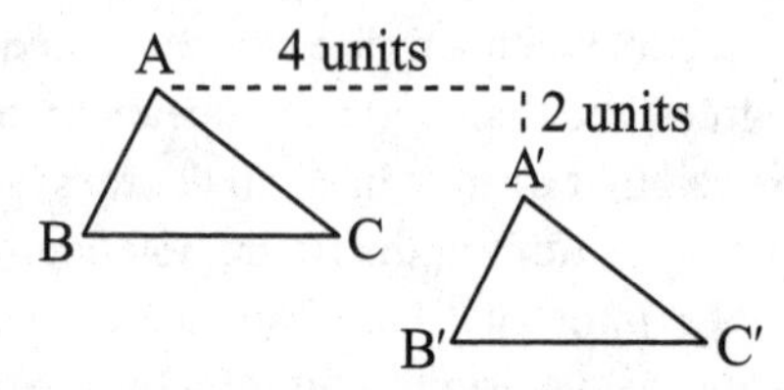

The symbol used to denote a translation is T. The translation of triangle ABC four units to the right and two units down is written as $\Delta ABC \xrightarrow{T_{180}} \Delta A'B'C'$.

Dilations

A figure is *dilated* if each part of the figure becomes larger or smaller depending on a given scale factor. Think of when the pupil of an eye is dilated at a physician's office. The pupil is still the same shape but the size of it chan ges.

Each time a dilation occurs, the change in size of the figure is based on a scale factor. For example, if the scale factor is 2, the new figure will be twice as large as the original figure. If the scale factor were $\frac{1}{3}$, the new figure will be one-third the size of the original figure.

The example below shows triangle ABC being dilated using a scale factor of 2 to become triangle *A'B'C'*. Notice that the shape is the same, but the image has sides that are twice as long as the original figure.

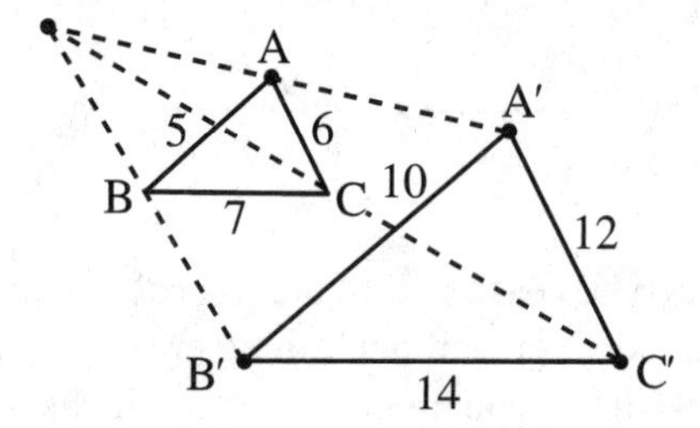

The symbol used to denote a dilation is D. The dilation of triangle ABC by a scale factor of 2 is written as $\Delta ABC \xrightarrow{D_2} \Delta A'B'C'$.

ISOMETRIES

An *isometry* is a transformation that does not change the size of a geometric figure. Line reflections, rotations, and translations are all isometries. Dilations of any scale factor other than 1 or −1 are not isometries; dilations of other scale factors change the size of the object.

STEPS YOU NEED TO REMEMBER

This section covers each of the transformations in the coordinate plane and the various types of problems associated with them.

Transformations in the Coordinate Plane

When dealing with transformations in the coordinate plane, there are specific locations for the figures and their images. This section will explain how to work with transformations within the coordinate system.

Line Reflections

Some of the most common line reflections in the coordinate plane are over the x-axis and the y-axis. In order to reflect a point over a line in the coordinate system, count the number of units the point is directly over from the line of reflection. The image of the point will be the same number of units on the opposite side of the line.

Take, for example, the point A(−2, 3). To reflect this point over the x-axis, count the number of units the point is directly above the x-axis. Since it is three units above the x-axis, count three units directly below the x-axis from this point to find the image of point A. This point is (−2, −3). Therefore, $A(-2,3)\xrightarrow{R_{x\text{-axis}}}A'(-2,-3)$. This is shown in the figure below.

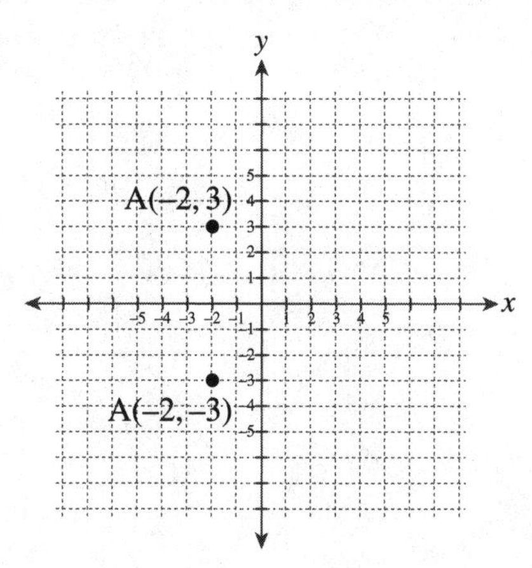

Thus, for any point $(x, y) \xrightarrow{R_{x\text{-axis}}} (x, -y)$.

Try now to reflect the same point over the y-axis. Again, start with the point A(–2, 3). To reflect this point over the y-axis, count the number of units the point is directly to the left of the y-axis. Since it is two units to the left of the y-axis, count two units directly to the right of the y-axis from this point to find the image of point A. This point is (2, 3). Therefore, $A(-2,3) \xrightarrow{R_{y\text{-axis}}} A'(2,3)$. This reflection is also shown in the figure below.

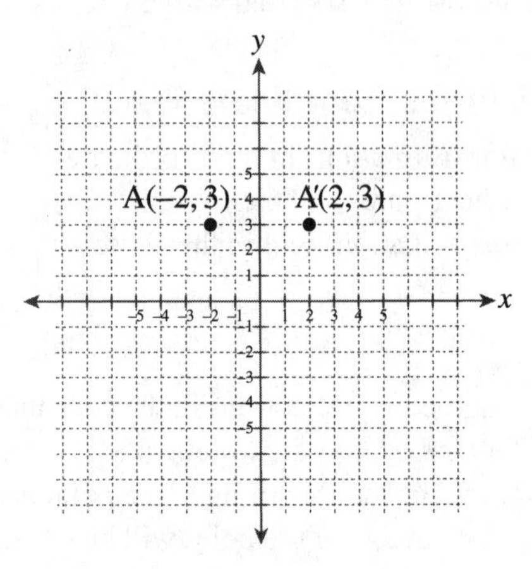

Thus, for any point $(x, y) \xrightarrow{R_{y\text{-axis}}} (-x, y)$.

The line y = x is also a common line of reflection. As discussed in Chapter 8, this line is a diagonal line that has a slope of 1 and a y-intercept of 0, so it crosses the y-axis at the origin. When reflecting over this line, you need to count the number of diagonal units from the line. Thus, the point B (3, 5) becomes B′ (5, 3), as shown in the figure below. In other words, the coordinates of the point switch when reflected over this line.

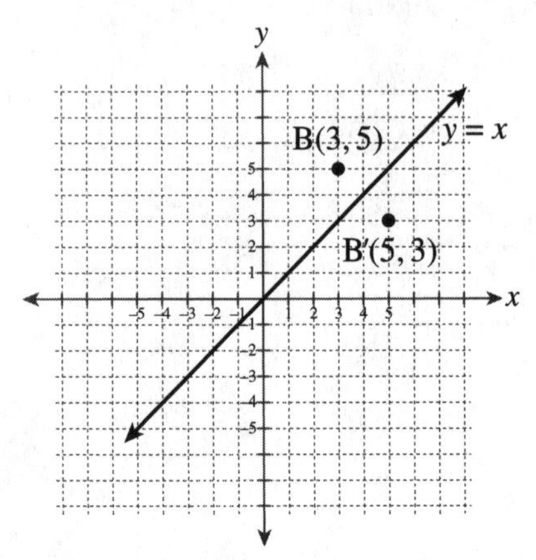

Thus, for any point $(x, y) \xrightarrow{R_{y=x}} (y, x)$.

Other common lines of reflection are vertical and horizontal lines. Recall from Chapter 8 that vertical lines are in the form $x = k$ and horizontal lines are in the form $y = k$. To reflect over these types of lines, first draw the line on coordinate axes. Then count the number of units the point is from the line of reflection to find the number of units the image of the point should be on the opposite side of the line. An example of reflecting the point M (4, 5) over the vertical line $x = 3$ to find the image M′ (2, 5) is shown in the figure below.

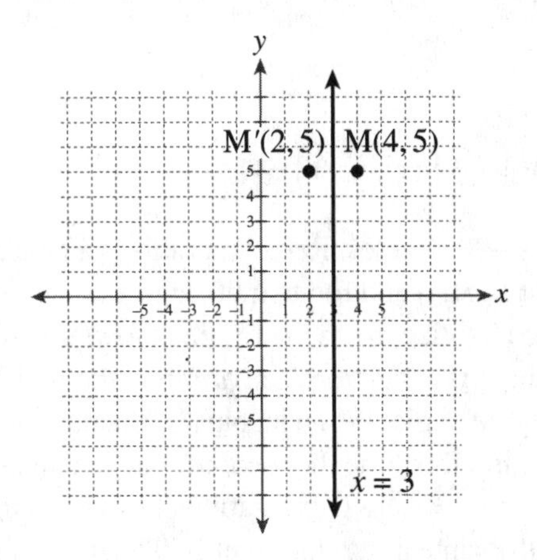

Rotations

When rotating geometric figures in the coordinate plane, the origin is the most common center of rotation. This means that the point being rotated moves around the origin much like the planets orbit the sun. The critical fact to know about a rotation is the degree of rotation. This is the number of degrees the point moves in a circle about the center.

Let's start with a rotation of 90 degrees. In the coordinate system, counterclockwise is a positive direction and clockwise is a negative direction, so this point is moving counterclockwise about the origin. If we take the point B (5, 4) and rotate it 90 degrees counterclockwise, the location of point B′ is (–4, 5). Notice, from the figure below, that if a line is drawn from the point to the origin and then the origin to the rotated point, it forms a 90-degree angle. When a point is rotated 90 degrees, the coordinates switch, and the x-coordinate becomes its opposite. So, $B(5, 4) \xrightarrow{ROT_{90}} B'(-4, 5)$. This rotation is shown in the figure below.

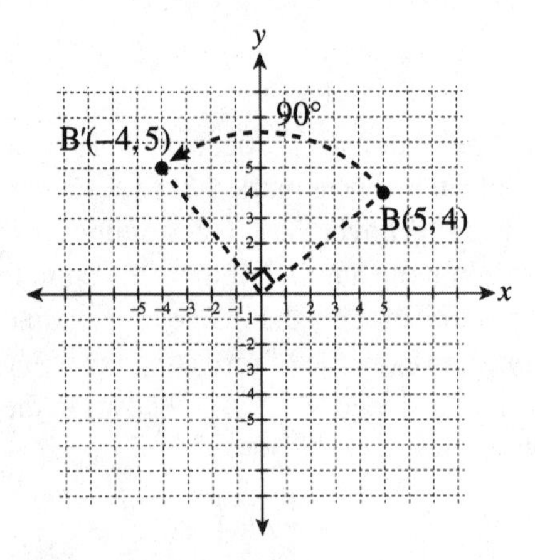

Thus, for any point $(x, y) \xrightarrow{ROT_{90}} (-y, x)$.

The next rotation is –90 degrees. Recall that this is a clockwise direction. If we take the point B(5, 4) and rotate it 90 degrees clockwise, the location of point B′ is (4, –5). Notice, from the figure below, that if a line is drawn from the point to the origin and then the origin to the rotated point, it also forms a 90-degree angle. When a point is rotated –90 degrees, the coordinates switch, but this time the *y*-coordinate becomes its opposite. So, $B(5, 4) \xrightarrow{ROT_{-90}} B'(4, -5)$. This rotation is shown in the figure below. Notice that this is the same as a rotation of 270 degrees in a counter-clockwise direction.

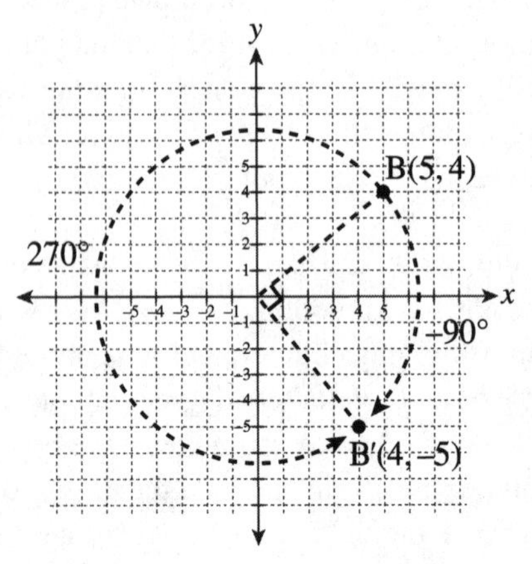

Thus, for any point $(x, y) \xrightarrow{ROT_{-90}} (y, -x)$.

A rotation of 180 degrees takes a point and moves it to a location on the opposite side of the origin from the original point. If we take the point B(5, 4) and rotate it 180 degrees, the location of point B′ is (–5, –4). Notice, from the figure below, that if a line is drawn from the point to the origin and then the origin to the rotated point, it forms a 180-degree, or straight, angle. When a point is rotated 180 degrees, each coordinate becomes its opposite. So, $B(5,4)\xrightarrow{ROT_{180}} B'(-5,-4)$. This rotation is shown in the figure below.

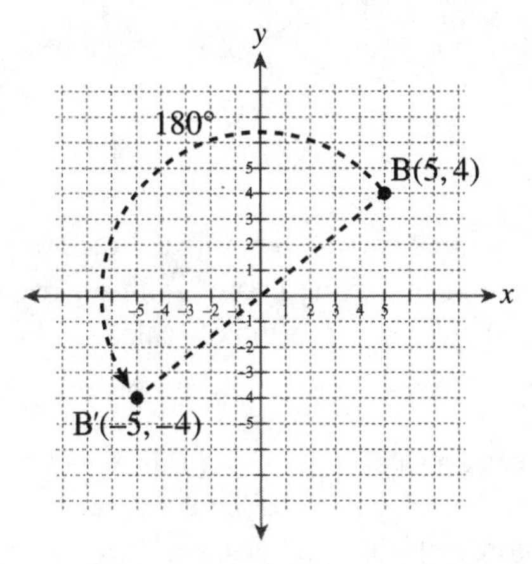

Thus, for any point $(x, y)\xrightarrow{ROT_{180}}(-x, -y)$.

Translations

A *translation* slides a point to a new location in the coordinate plane. A translation is written using the symbol $T_{a,b}$. A translation that moves a point *a* units over and *b* units up or down is written as $(x, y)\xrightarrow{T_{a,b}}(x+a, y+b)$. Each point in a translation is moved *a* units to the right if the value of *a* is positive or to the left if the value of *a* is negative. In a similar fashion, each point is also moved *b* units up if the value of *b* is positive or *b* units down if the value of *b* is negative.

In order to find the coordinates of the image after a translation, add *a* to the *x*-coordinate and *b* to the *y*-coordinate. For example, take the point C(–2, 6). If the point is translated $T_{3,-1}$ this will move the point three units to the right and one unit down from C. The image is $C(-2,6)\xrightarrow{T_{3,-1}} C'(-2+3, 6+-1)$, which is the point C′(1, 5). An example of the translation of point C is shown in the figure below.

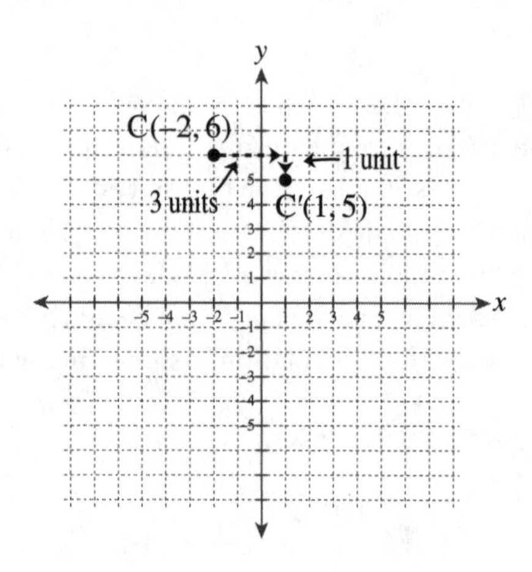

Dilations

A *dilation* either shrinks or enlarges an object, based on the scale factor. To find the image of a dilated point, multiply each coordinate by the scale factor. Therefore, a dilation of scale factor *a* is written as $(x, y) \xrightarrow{D_a} (ax, ay)$.

For example, take triangle DEF with point D (−3, −2), point E (−1, −2) and point F (−2, −5). If each point is dilated with a scale factor of two, this will double each coordinate. The images are $D(-3,-2) \xrightarrow{D_2} D'(-6,-4)$, $E(-1,-2) \xrightarrow{D_2} E'(-2,-4)$, and $F(-2,-5) \xrightarrow{D_2} F'(-4,-10)$. An example of the dilation of triangle DEF is shown in the figure below.

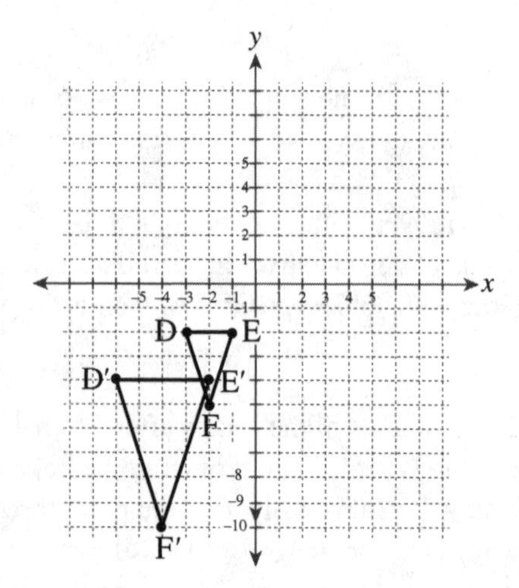

You may also be presented with dilations with negative scale factors. In these cases, the scale factor is still multiplied by the coordinates of each point. However, the image of the figure will also be rotated 180 degrees as well as dilated because each of the coordinates will also change signs. Take the example in the figure below. Triangle XYZ was dilated with a scale factor of –2. Notice that triangle X′Y′Z′ is twice as large as the original triangle but has also been rotated 180 degrees about the origin.

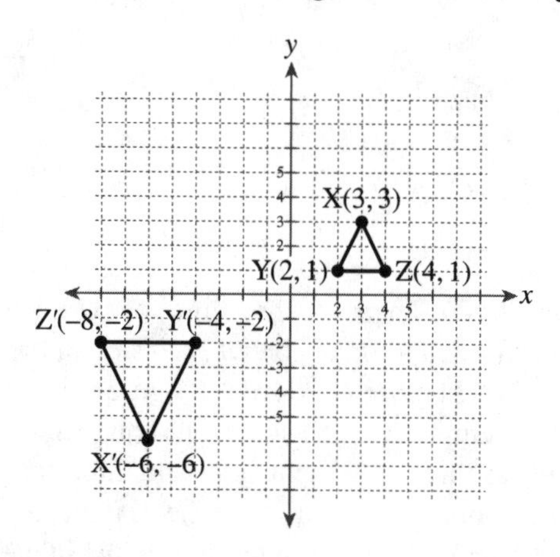

SUMMARY OF THE RULES FOR TRANSFORMATIONS IN THE COORDINATE PLANE

Reflections

$(x, y) \xrightarrow{R_{x-axis}} (x, -y)$

$(x, y) \xrightarrow{R_{y-axis}} (-x, y)$

$(x, y) \xrightarrow{R_{y=x}} (y, x)$

Rotations

$(x, y) \xrightarrow{ROT_{90}} (-y, x)$

$(x, y) \xrightarrow{ROT_{-90}} (y, -x)$

$(x, y) \xrightarrow{ROT_{180}} (-x, -y)$

Translations

$(x, y) \xrightarrow{T_{a,b}} (x + a, y + b)$

Dilations

$(x, y) \xrightarrow{D_a} (ax, ay)$

STEP-BY-STEP ILLUSTRATION OF THE FIVE MOST COMMON QUESTION TYPES

Question 1: Line Symmetry

How many lines of symmetry are there in an equilateral triangle?

(A) 0

(B) 1

(C) 2

(D) 3

(E) 6

The correct choice is (D). There are three lines of symmetry. A line of symmetry is drawn through a figure so that each side of the line is a mirror image of the other side. In other words, if the figure was drawn on paper and folded on the line of symmetry, each side would map onto the other side. An equilateral triangle has three congruent sides and three congruent angles. A line can be drawn that bisects each angle of the triangle, and this line will be a line of symmetry. The lines of symmetry are shown in the figure below.

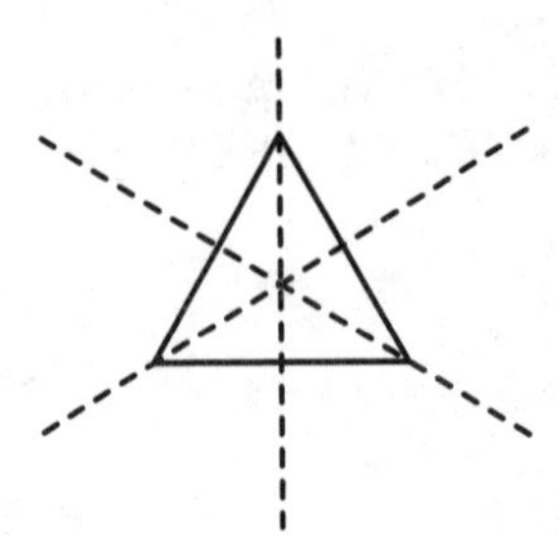

Question 2: Line Reflections

What are the coordinates of the point (-4, 3) after a reflection over the *y*-axis?

(A) (4, 3)

(B) (3, −4)

(C) (−3, −4)

(D) (−4, −3)

(E) (4, −3)

The correct choice is (A). As discussed earlier in this chapter, a reflection is a mirror image of a figure. When a point is reflected, it will be the same distance away from the line of reflection just on the opposite side of the line. The point (–4, 3) is located 4 units to the left of the origin and 3 units up in quadrant II. In order to reflect this point over the y-axis, count the number of units it is from the y-axis. Since it is 4 units to the left, count 4 units directly over to the right of the y-axis to find the reflected point. This is the point (4, 3). This reflection is shown in the figure below.

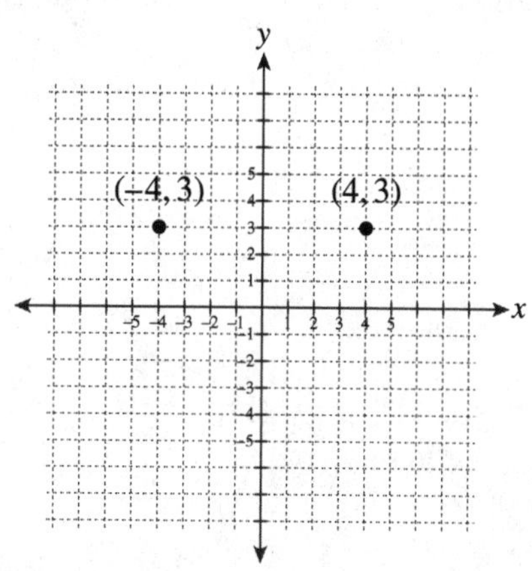

Another way to find the coordinates of the reflected point is to use the rule for reflecting over the y-axis. This rule was stated as $(x, y)\xrightarrow{Ry\text{-}axis}(-x, y)$. This rule says that the x-coordinate becomes its opposite when a point is reflected over the y-axis. Thus, the point (–4, 3) becomes (4, 3).

Choice (B) is the result of reflecting the point over the line $y = x$. Choice (C) is the result of rotating the point 90 degrees. Choice (D) is the result of reflecting the point over the x-axis. Choice (E) is the result of rotating the point 180 degrees.

Question 3: Rotations

What are the coordinates of the point (2, –5) after a rotation of 90 degrees?

(A) (–5, 2)

(B) (5, 2)

(C) (5, –2)

(D) (–5, –2)

(E) (–2, 5)

The correct answer choice is (B). In the coordinate system, counterclockwise is a positive direction and clockwise is a negative direction, so this point is moving counterclockwise about the origin. Take the point (2, –5) and rotate it 90 degrees counterclockwise, the location of the rotated point is (5, 2). Notice, from the figure below, that if a line is drawn from the point to the origin and then the origin to the rotated point, it forms a 90-degree angle.

Another way to find the coordinates of the rotated point is to use the formula presented earlier in this chapter. When a point is rotated 90 degrees, the coordinates switch, and the *x*-coordinate becomes its opposite. So, $(2,-5)\xrightarrow{ROT_{90}}(5,2)$. This rotation is shown in the figure below.

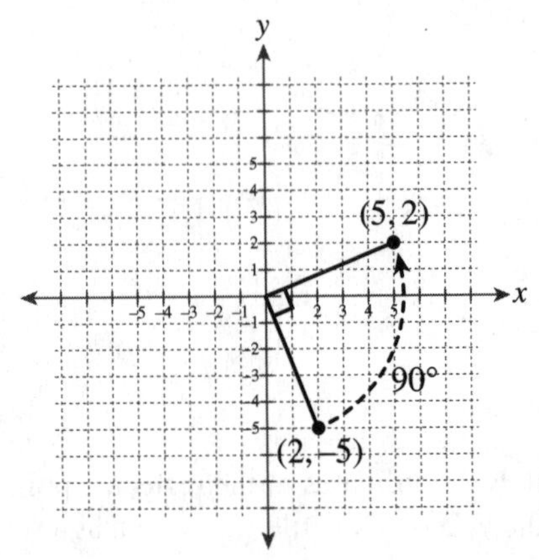

Choice (A) is the coordinates of the point after a reflection over the line $y = x$. Choice (E) is the result of rotating the given point 180 degrees.

Question 4: Translations

The coordinates of point A after a translation of $T_{2,-3}$ is (0, 6). What are the coordinates of point A?

(A) (0, –18)

(B) (–2, 3)

(C) (–2, 9)

(D) (–2, –9)

(E) (2, 3)

The correct answer is (C). A translation that moves a point *a* units over and *b* units up or down is written as $(x, y)\xrightarrow{T_{a,b}}(x+a, y+b)$. Each point in a translation is moved *a* units to the right if the value of *a* is positive or to the left if the value of *a* is negative. In a similar fashion, each point is also moved *b* units up if the value of *b* is positive or *b* units down if the value of *b* is negative.

In this question, however, you are given the coordinates of the image of the point and asked to find the original point. A strategy that can be used is to work backwards. In order to find the coordinates of point A, start by using the formula. The formula would be $A(x, y) \xrightarrow{T_{2,-3}} A'(0, 6)$. Since a translation of $T_{2,-3}$ moves the point 2 units to the right and 3 units down, move 3 units up and 2 units to the left from A′. This is the point (-2, 9). Therefore, $A(-2, 9) \xrightarrow{T_{2,-3}} A'(0, 6)$. This translation is shown in the figure below.

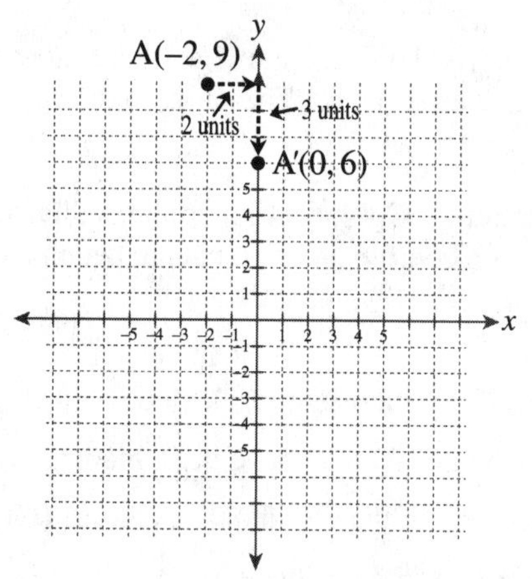

You may have selected choice (A) if you had multiplied the coordinates by the values in the translation. Choice (B) is the result of subtracting the first value and adding the second value of the translation from the given point. Choice (D) is the result of a mathematical error. Choice (E) is the result of using point A′ as the original point, and not using it as the image of A.

Question 5: Dilations

Given triangle ABC in the diagram below, what are the coordinates of point C after a dilation of scale factor 3?

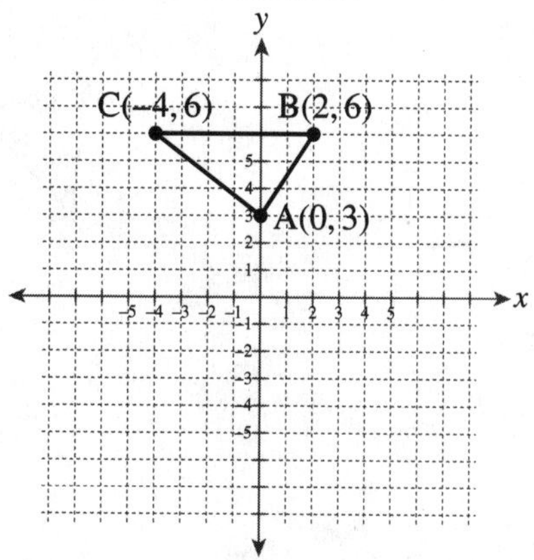

(A) (−1, 9)

(B) (−7, 3)

(C) $\left(\frac{-4}{3}, 2\right)$

(D) (12, −18)

(E) (−12, 18)

The correct answer is (E). To find the image of a dilated point, multiply each coordinate by the scale factor. Therefore, a dilation of scale factor a is written as $(x, y) \xrightarrow{D_a} (ax, ay)$.

The point in this question is C(−4, 6). If the point is dilated with a scale factor of 3, this will triple each coordinate of point C to become C′. The image is $C(-4, 6) \xrightarrow{T_{2,-3}} C'(-12, 8)$.

Choice (A) may have been incorrectly selected if 3 were added to each of the coordinates. Choice (B) may have been incorrectly selected if 3 was subtracted from each of the coordinates. Choice (C) is the result of incorrectly multiplying each coordinate by $\frac{1}{3}$. Choice (D) is the result of incorrectly multiplying each coordinate by −3, instead of 3.

CHAPTER QUIZ

1. How many lines of symmetry are there in a block letter H?

   (A) 0

   (B) 1

   (C) 2

   (D) 3

   (E) 4

2. Which of the following represents a line reflection?

   (A)

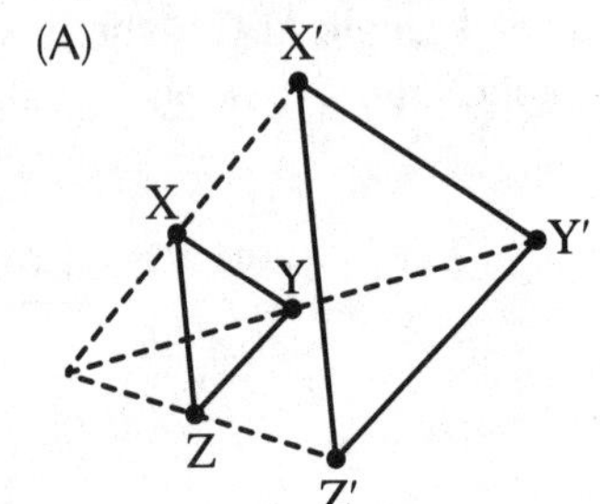

(B)

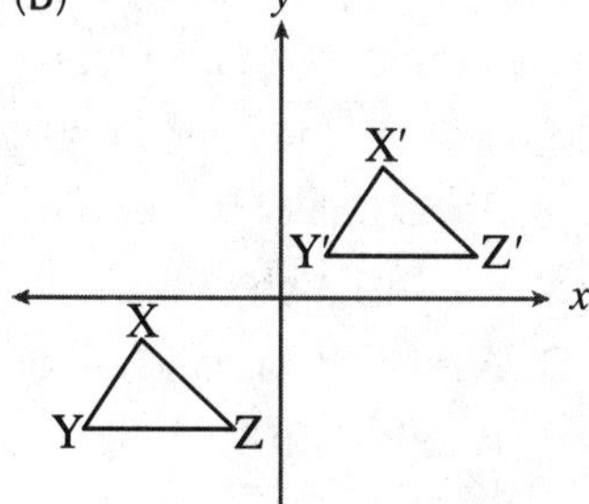

(C)

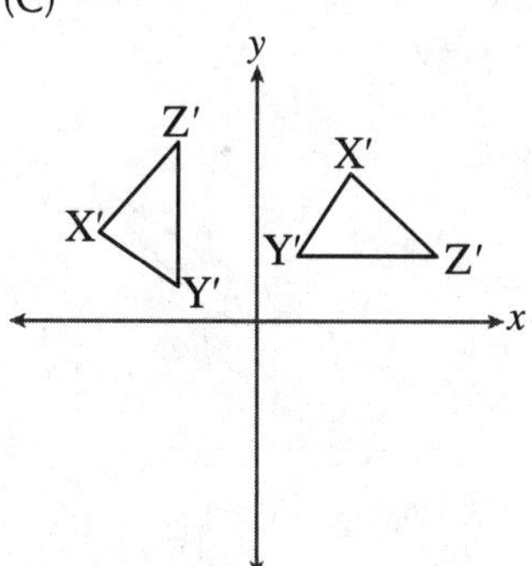

(D)

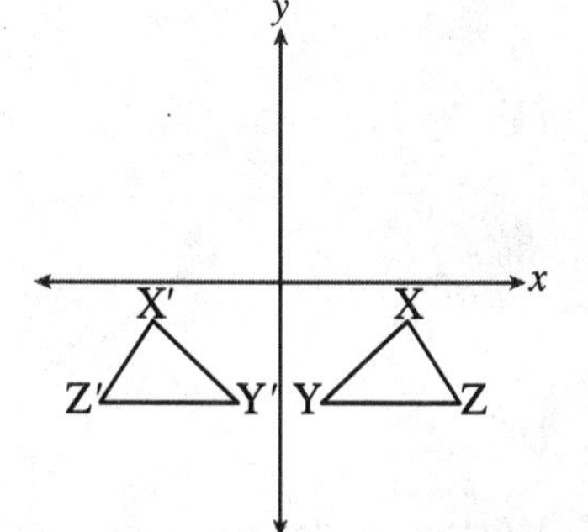

(E)

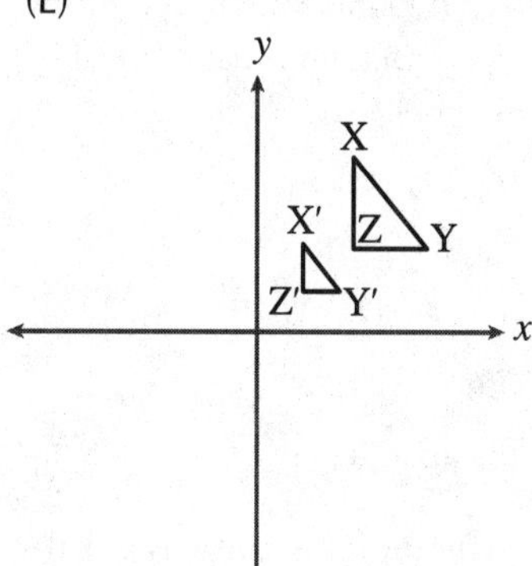

3. The point is the image of point G′ after a reflection over which of the following lines?

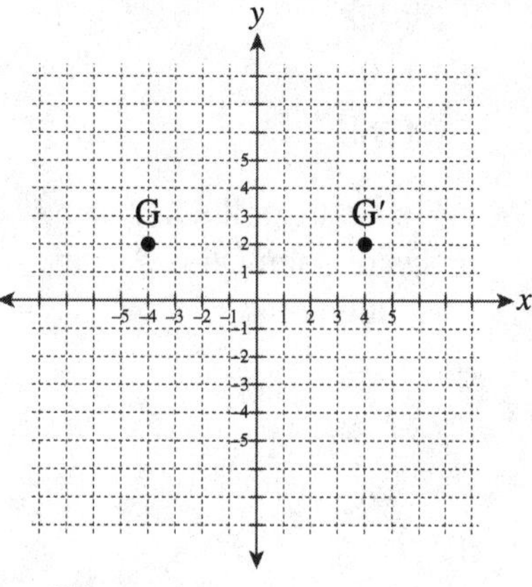

(A) y-axis

(B) x-axis

(C) $y = x$

(D) $y = 2$

(E) $x = 4$

4. What is the image of the point (−4, −6) after a rotation of −90 degrees?
   (A) (4, 6)
   (B) (−6, −4)
   (C) (6, 4)
   (D) (−6, 4)
   (E) (6, −4)

5. Which of the following is the number of degrees of rotational symmetry in a square?
   (A) 30°
   (B) 60°
   (C) 90°
   (D) 120°
   (E) 150°

6. Which of the following is the image of point D(−6, 7) after a translation of $T_{9,-3}$?
   (A) (9, −3)
   (B) (3, 4)
   (C) (15, 10)
   (D) (6, −7)
   (E) (−54, −21)

7. The coordinates of point P′ after a translation of $T_{-5,-2}$ is (2, 2). What are the coordinates of point P?
   (A) (−10, −4)
   (B) (−7, 0)
   (C) (7, −4)
   (D) (−3, 0)
   (E) (7, 4)

8. Which of the following is **not** an example of an isometry?
   (A) A reflection over the y-axis
   (B) A rotation of 90 degrees
   (C) A translation of T-3,-5
   (D) A dilation of scale factor 3
   (E) A reflection over the line $y = x$

9. Which of the following is the scale factor used in the dilation below?

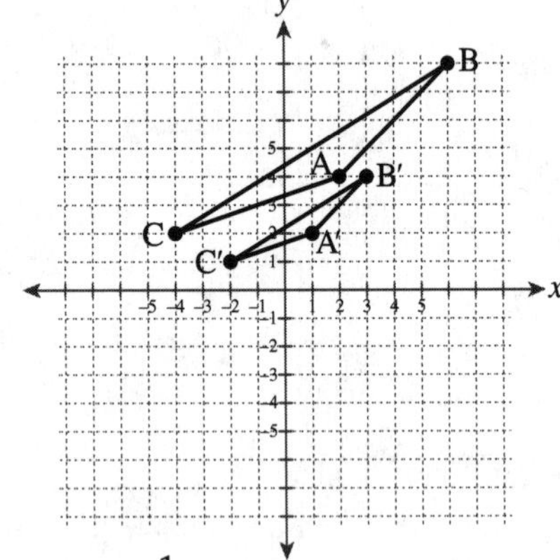

   (A) $\frac{1}{2}$
   (B) $-\frac{1}{2}$
   (C) 1
   (D) 2
   (E) −2

10. What is the image of the point A(5, −15) after a dilation scale factor 5?
   (A) (0, −20)
   (B) (1, −3)
   (C) (5, −10)
   (D) (10, −10)
   (E) (25, −75)

ANSWER EXPLANATIONS

1. C

There are two lines of symmetry. A line of symmetry is drawn through a figure so that each side of the line is a mirror image of the other side. Therefore, if the figure was drawn on paper and folded on the line of symmetry, each side would map onto the other side. A block letter H has vertical line symmetry and horizontal line symmetry. Each side would be a mirror image of the other side when a vertical line or a horizontal line is drawn. The lines of symmetry are shown as dotted lines in the figure below.

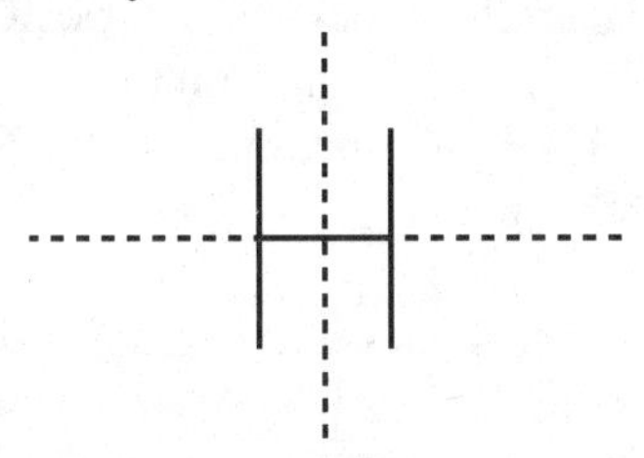

2. D

A line reflection is a transformation that reflects, or flips, an object over a line. The image of the object is the same distance away from the line of reflection as the original object and appears as a mirror image of the original figure. This occurs in choice (D).

Choice (A) is a dilation of scale factor 2. Choice (B) is an example of a translation, or slide. Choice (C) is a rotation of 90 degrees. Choice (E) is a dilation of scale factor $\frac{1}{2}$.

3. A

A reflection is a mirror image of a figure. When a point is reflected, the image will be perpendicular to and the same distance away from the line of reflection, just on the opposite side of the line. The points G and G′ are each located four units away from the y-axis. G is four units to the left of the y-axis at the point (–4, 2) and point G′ is four units to the right of the y-axis at the point (4, 2). Thus, the line of reflection is the y-axis.

Another way to find the line of reflection is to use the rule for reflecting over the y-axis. This rule was stated as $(-4, -6) \xrightarrow{ROT_{-90}} (-6, 4$. This rule says that the x-coordinate becomes its opposite when a point is reflected over the y-axis. Since the point G(–4, 2) becomes (4, 2), the point was reflected over the y-axis.

For choice (B), each point would be the same distance from the x-axis and the y-coordinate would change sign. For choice (C), each point would be the same distance from the line $y = x$. This is a diagonal line with slope equal to 1 and a y-intercept of zero. When a point is reflected over the line $y = x$, the x- and y-coordinates switch, which is not what happened in the figure. The line of reflection in choice (D) is a horizontal line that would go through each of the indicated points and would not be the line of reflection. Choice (E) is a vertical line that would go through point G′, but would not be the line of reflection.

4. D

In the coordinate system, counterclockwise is a positive direction and clockwise is a negative direction, so this point is moving clockwise about the origin. Take the point (–4, –6) and rotate it 90 degrees clockwise, and the location of the rotated point is (–6, 4).

Another way to find the coordinates of the rotated point is to use the formula presented earlier in this chapter. When a point is rotated –90 degrees, the coordinates switch, and the y-coordinate becomes its opposite. So, $(-4, -6) \xrightarrow{ROT_{-90}} (-6, 4)$.

Choice (A) is the coordinates of the point after a rotation of 180 degrees. Choice (B) is the result of reflecting the point over the line $y = x$. Choice (E) is the result of rotating the point 90 degrees counterclockwise in a positive direction.

5. C

Rotational symmetry occurs when a figure can be rotated, or turned, a certain number of degrees less than 360 and appears the same as the original figure. The number of degrees of rotational symmetry is based on the number of times you see an image that is the same as the original figure. In a square, there are four congruent sides and four congruent angles. Each angle is 90 degrees in measure. If a square is rotated, you will see an image that appears the same as the original figure at four different times. Therefore, a square has a 90 degree rotational symmetry, since $360 \div 4 = 90$.

6. B

A translation slides a point to a new location in the coordinate plane. A translation is written using the symbol $T_{a,b}$. A translation that moves a point a units over and b units up or down is written as $(x, y) \xrightarrow{T_{a,b}} (x + a, y + b)$. Each point in a translation is moved a units to the right if the value of a is positive or to the left if the value of a is negative. In the same way, each point is also moved b units up if the value of b is positive or b units down if the value of b is negative.

In order to find the coordinates of the image after a translation, add a to the x-coordinate and b to the y-coordinate. For this problem, take the point D(–6, 7). If the point is translated $T_{9,-3}$ this will move the point nine units to the right and three units down from D. The image is $D(-6,7) \xrightarrow{T_{9,-3}} D'(-6+9, 7+-3)$, which is the point D′ (3, 4).

Choice (A) is the numbers in the translation, not the image of the point. Choice (C) is the result of adding the coordinates of point D with the values of the translation, without regard to the negative signs. Choice (E) is the result of multiplying the coordinates by the values of the translation, instead of adding them.

7. E

A translation that moves a point a units over and b units up or down is written as $(x, y) \xrightarrow{T_{a,b}} (x+a, y+b)$. Each point in a translation is moved a units to the right if the value of a is positive or to the left if the value of a is negative. In the same way, each point is also moved b units up if the value of b is positive or b units down if the value of b is negative.

In this question, however, you are given the coordinates of the image of the point and asked to find the original point. Use the strategy of working backwards. In order to find the coordinates of point P, start by using the formula. The formula would be $P(x, y) \xrightarrow{T_{-5,-2}} P'(2, 2)$. Since a translation of $T_{-5,-2}$ moves the point 5 units to the left and 2 units down, move 2 units up and 5 units to the right from P′. This is the point (7, 4). Therefore, $P(7, 4) \xrightarrow{T_{-5,-2}} P'(2, 2)$.

You may have selected choice (A) if you had multiplied the coordinates by the values in the translation. Choice (B) is the result of subtracting the first value and adding the second value of the translation from the given point. Choice (C) is the result of a mathematical error. Choice (D) is the result of using point P′ as the original point, and not using it as the image of P.

8. D

An isometry is a transformation that does not change the size of a geometric figure. Line reflections, rotations, and translations are all isometries. Dilations of any scale factor other than 1 or –1 are not isometries; dilations of other scale factors change the size of the object. For the choices in this question, each of the choices is an isometry except for a dilation of scale factor 3. This will transform the given object to be three times the size of the original object. Each of the other answer choices will not change the size of the original figure.

9. A

A dilation changes an object by a scale factor. This scale factor is multiplied by each of the coordinates of the points that make up the object in order to form the image of the object.

Compare each of the corresponding coordinates of triangle ABC:

Point A (2, 4) becomes point A′ (1, 2).

Point B (6, 8) becomes point B′ (3, 4).

Point C (–4, 2) becomes point C′ (–2, 1).

The coordinates of the image of each point is equal to one-half of the original coordinates. Therefore, the dilation scale factor is $\frac{1}{2}$.

In choice (B), a scale factor of $-\frac{1}{2}$ is given. This would cause the image of the original figure to be half the size, but would also rotate the figure 180 degrees about the origin.

In choice (C), a scale factor of 1 would not change the triangle. Choice (D) would have made the image twice as large as the original. Choice (E) would also make the image twice as large and also rotate the image 180 degrees about the origin.

10. E

To find the image of a dilated point, multiply each coordinate by the scale factor. Therefore, a dilation of scale factor *a* is written as $(x, y) \xrightarrow{Da} (ax, ay)$.

The point in this question is A (5, –15). If the point is dilated with a scale factor of 5, this will multiply each coordinate of point A by five to become A′. The image is $A(5, -15) \xrightarrow{D_5} A'(25, -75)$.

Choice (A) may have been incorrectly selected if 5 were subtracted from each of the coordinates. Choice (B) may have been incorrectly selected if each of the coordinates were divided by five. Choice (C) is the result of incorrectly adding five to the *y*-coordinate only. Choice (D) is the result of incorrectly adding five to each coordinate.

CHAPTER 11

Locus of Points

WHAT IS A LOCUS OF POINTS?

A *locus of points* is a set of points that fit a certain set of criteria. Common locus of points problems deal with the set of points equidistant from a point, two points, a line, and two lines. In addition, there are also situations where more than one set of criteria is met by a set of points; this is called *compound loci*. In this chapter, you will learn each of the basic conditions of locus and review the formulas and equations that are used when finding a locus of points.

CONCEPTS TO HELP YOU

This section will discuss the five basic conditions of locus of points. Sometimes a locus of points can be viewed as a force field around an object, as you will see in the diagrams in this section. The concept of compound loci and the points that satisfy more than one condition at a time will also be explained in this section.

The Five Basic Conditions of Locus

There are five basic conditions of a locus of points: the locus of points (1) about a single point, (2) from two points, (3) from one line, (4) from two parallel lines, and (5) from two intersecting lines. Each of these conditions is explained in this section.

Locus about a Point

The locus of points about a single point is a circle that surrounds the point. The radius of the circle is the number of units each point is away from the points. For example, the locus of points 4 units from point P is a circle surrounding point P with a radius of 4 units. This locus is shown in the figure below as a dotted circle around point P.

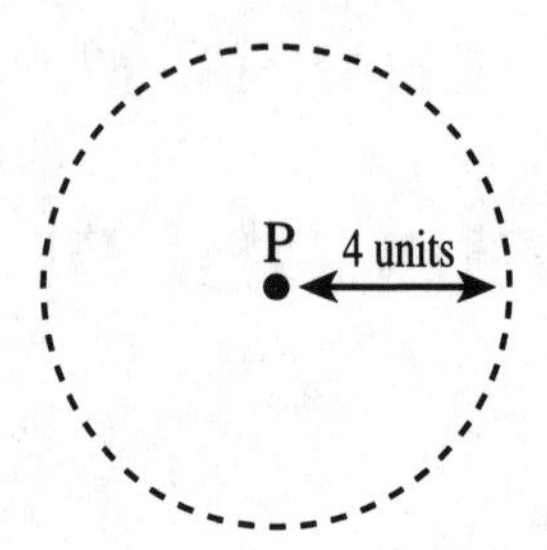

Locus from Two Points

The locus of points from two given points is a line that is located between the two points. This line is created by the set of points equidistant from both points at the same time. The locus is the perpendicular bisector of the line segment that would connect the two points. An example of the locus of points equidistant from points A and B is shown in the figure below as a dotted line.

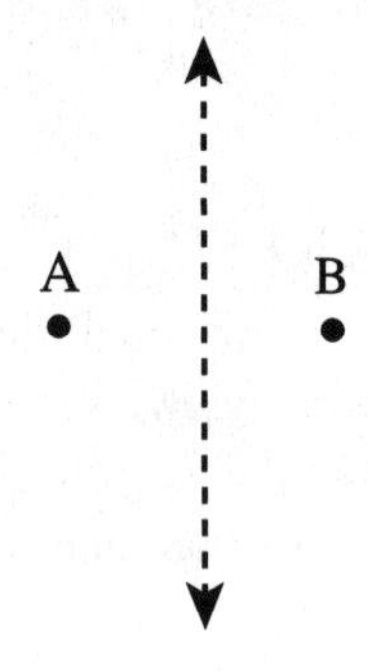

Locus about a Line

The locus of points about a line consists of two parallel lines on either side of the line, each the same distance from the line. In the figure below, the locus of points from line l is line m and line n. Each is shown as dotted lines in the figure. Each point on lines m and n is the same distance from line l. Each of these lines is parallel to line l and will never intersect it.

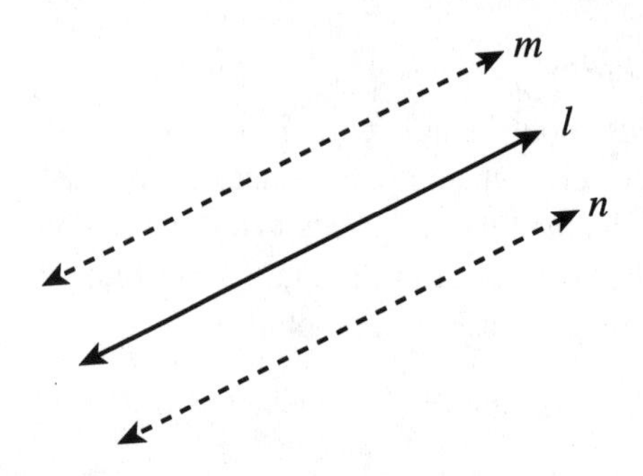

Locus from Two Parallel Lines

The locus of points from two parallel lines is a line that is located between the two lines that are also parallel to the lines. Since the points contained in the locus are equidistant from the two given lines, this line will be located the same distance from each given line. In the figure below, lines p and q are given. The locus of points equidistant from these two lines is the dotted line located in between these two lines the same distance from each line. This line is also parallel to lines p and q.

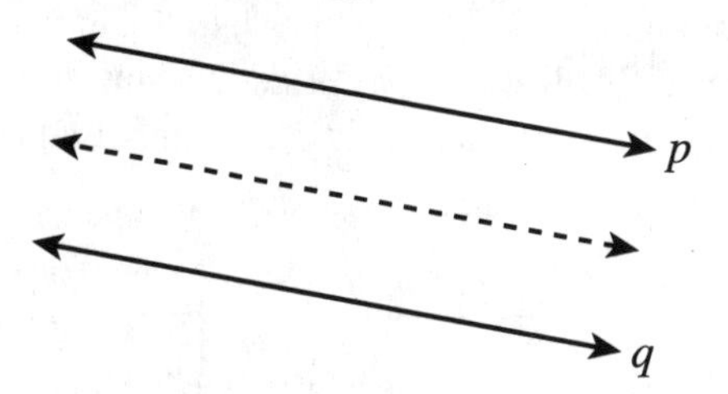

Locus from Two Intersecting Lines

The locus of points from two intersecting lines is a set of points that forms two other intersecting lines. The lines that form the locus of points bisect the angles that are formed by the two given lines. In the figure below, line r and line s intersect at point A. The locus of points equidistant from each of these lines is the dotted lines that also intersect at point A. These dotted lines bisect the vertical angles that are formed by line r and line s. Recall from Chapter 1 that vertical angles are the nonadjacent angles formed when any two lines intersect.

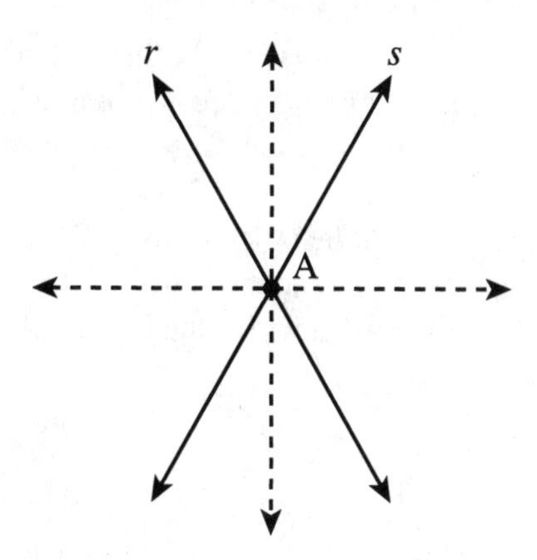

Compound Loci

Sometimes there is more than one condition to be met when finding the locus of points. If the points satisfy more than one set of criteria, it is called *compound loci.* A common example of compound loci is the set of points a certain distance from a given point and the set of points a certain distance from a given line. Take the figure below. This figure shows the number of points 6 units from line l **and** 4 units from a given point P that is not on line l. The dotted line surrounding point P is the circle that forms the locus of points 4 units from point P. The dotted lines parallel to line l form the locus of points 6 units from that line. These dotted lines intersect in two places; therefore, there are two points that satisfy the compound loci.

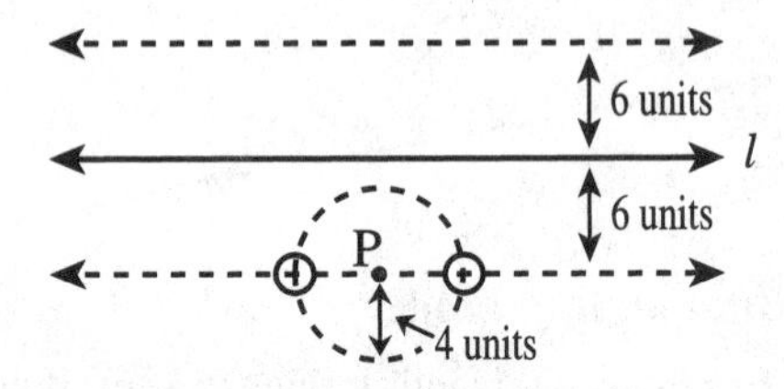

STEPS YOU NEED TO REMEMBER

Finding the locus of points from various objects in the coordinate plane is a common application of this topic. This section will describe the process of using a coordinate system with locus of points and finding a specific location and/or equation for each locus.

Basic Locus in the Coordinate Plane

When the locus of points is found in the coordinate plane, an exact location of the locus can be found. Start with the locus of points equidistant from a given line. The locus of points one unit from the x-axis is a set of two horizontal parallel lines. One line is located one unit above the x-axis, and the other line is located one unit below the x-axis. The equation of the line one unit above the x-axis is $y = 1$, and the equation of the line one unit below the x-axis is $y = -1$. The locus is shown in the figure below.

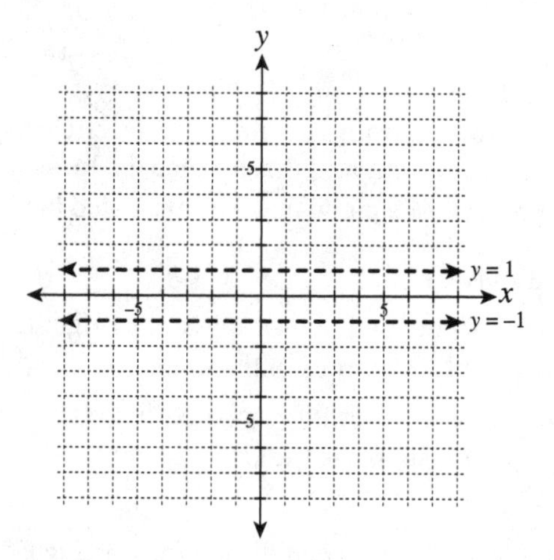

Finding the equation of various types of lines was discussed in Chapter 8, so refer to that chapter for help on this topic.

Another type of basic locus problem in the coordinate plane is finding the equation of the locus of points equidistant from two given lines. For this example, let's use the lines $x = 3$ and $x = 7$. Each of these equations represents a vertical line. To find the locus, find the set of points the same distance from both of the lines at the same time. In this case, the lines are four units apart, so the locus of points will be two units from each line at a location between the two given lines. This set of points will form a line parallel to the two given lines. The equation of this line is $x = 5$, and is shown as a dotted line in the figure below.

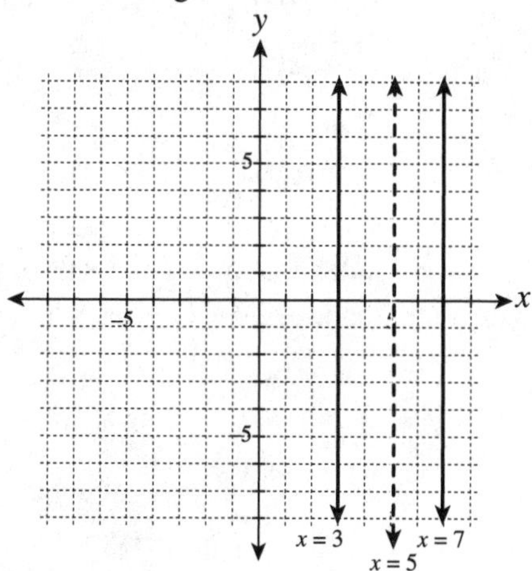

Equation of a Circle

The set of points equidistant from a single point is a circle that surrounds the points. The radius of the circle is the distance the locus is from the given point. Recall the equation of a circle that was discussed in Chapter 9.

EQUATION OF A CIRCLE

The equation of a circle with the center at the origin is $x^2 + y^2 = r^2$, where r represents the radius of the circle.

The equation of a circle with the center at (h, k) is $(x - h)^2 + (y - k)^2 = r^2$, where r represents the radius of the circle.

When locating the locus of points equidistant from a single point in the coordinate plane, use the above formulas to write the equation. For example, the locus of points 5 units from the origin would have the equation $x^2 + y^2 = 25$. This locus of points is shown in the figure below.

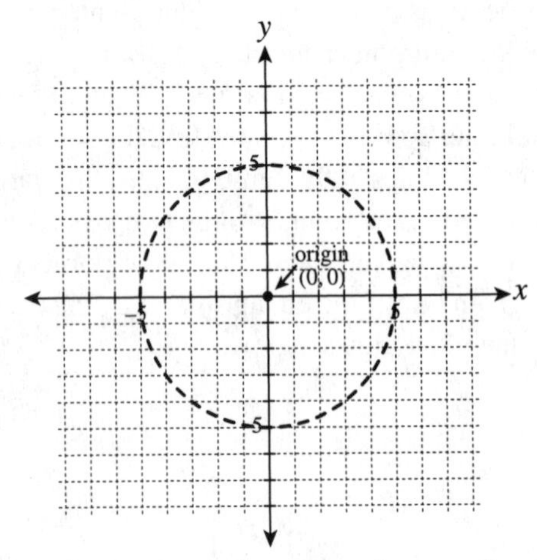

If the given point is not located at the origin, the equation is slightly different as highlighted in the sidebar above. The locus of point 3 units from the point $(-2, 1)$ has the equation $(x + 2)^2 + (y - 1)^2 = 9$, and is shown on the next page.

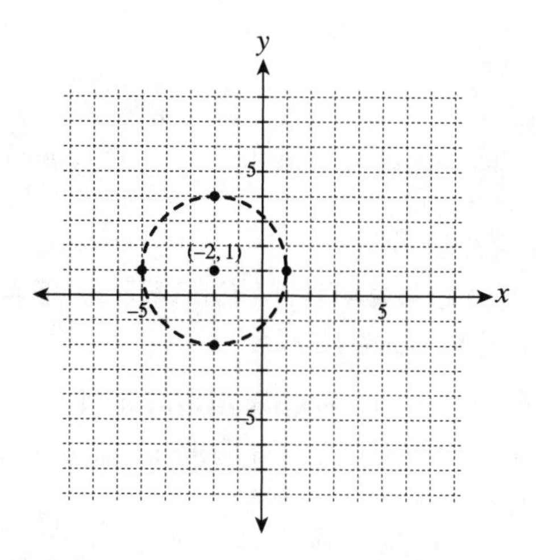

Compound Loci in the Coordinate Plane

Compound loci problems also appear in the coordinate plane. Take the following example.

> How many points are contained in the locus of points 2 units from the origin and equidistant from the lines $y = 1$ and $y = -3$?

In order to solve this problem, first graph the locus of points for each condition. Then, find the number of places where the loci intersect. This will be the total number of points that satisfy both conditions at the same time.

The locus of points 2 units from the origin is a circle that surrounds the origin with center at the origin, or (0, 0). The equations $y = 1$ and $y = -3$ represent two horizontal lines. The locus of points equidistant from two lines is a line parallel to these two lines contained halfway between the lines. The equation of the locus of points is $y = -1$. The locus of points for each of these conditions is shown as dotted lines in the figure below.

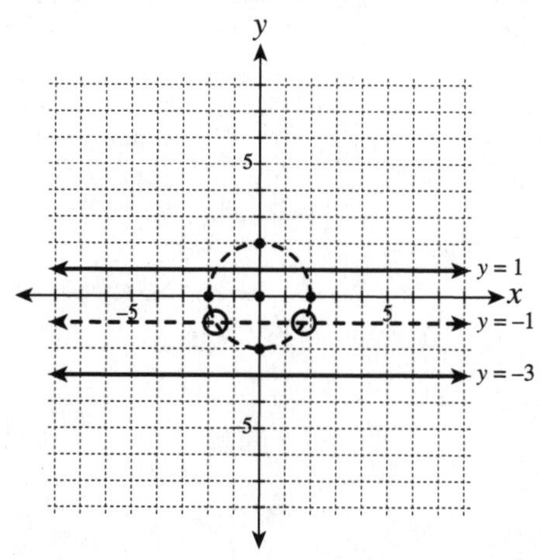

The dotted lines intersect in two locations. Each is circled in the above figure. Therefore, there are two points that satisfy the compound loci stated in this question.

STEP-BY-STEP ILLUSTRATION OF THE FIVE MOST COMMON QUESTION TYPES

Question 1: The Locus of Points About a Line

Which of the following best describes the locus of points about a horizontal line *m*?

(A) A single point located on line *m*

(B) A line located above m and parallel to *m*

(C) A line located below m and parallel to *m*

(D) A line perpendicular to line *m*

(E) A line above and a line below line *m*, both parallel to line *m*

The correct answer is (E). The locus of points is the set of points equidistant from an object(s) that meet a certain set of criteria. The set of points equidistant from a given line form two lines parallel to the given line and are located on either side of the line. The line in this question is a horizontal line, so the lines that form the locus of points will be horizontal and located both above and below the given line. A possible diagram showing the locus of points equidistant from a given horizontal line *m* is shown in the figure below.

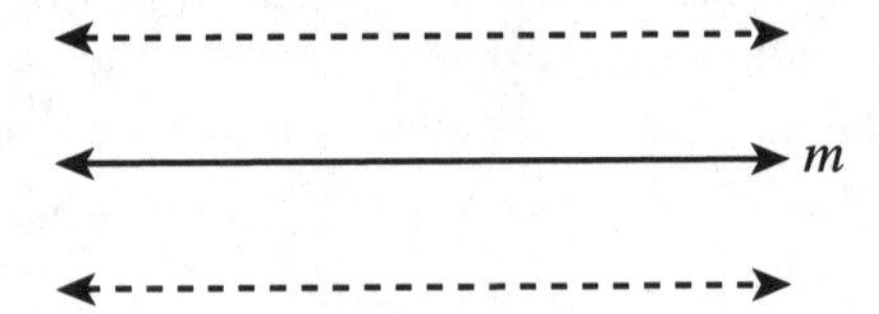

Choices (B) and (C) are each part of the locus, but neither describe the entire locus of points.

Question 2: The Locus of Points About Two Points

Point A is (3, 2) and point B is (5, 2). What is the equation of the locus of points equidistant from these two points?

(A) $y = 2$

(B) $x = 4$

(C) $x = 2$

(D) $y = 4$

(E) $x = 8$

The correct answer is choice (B). To find the locus of points, first find the location of the two given points on a coordinate grid. Each is located in the first quadrant and is shown in the figure below. The locus of points equidistant from two given points is a line between the two points. This line would be the perpendicular bisector of the line segment that would connect them. The locus is shown as a dotted line in the figure below. The equation of this line is in the form $x = k$, since it is a vertical line. The equation of the line is $x = 4$, since 4 is the halfway point between 3 and 5.

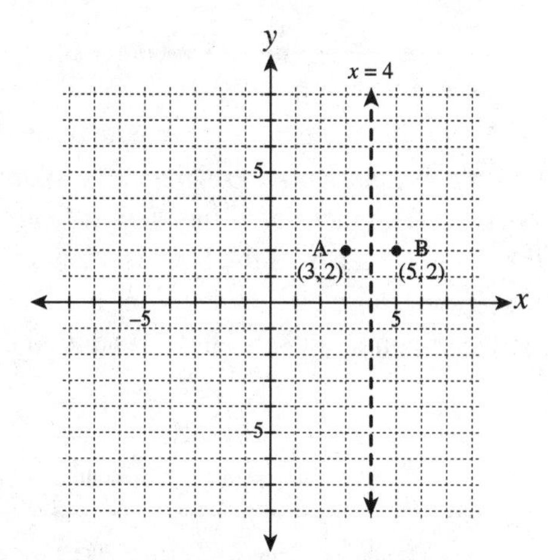

Choice (A) is a horizontal line two units above the x-axis, and choice (D) is a horizontal line four units above the x-axis. Neither of these lines would be the perpendicular bisector of the line segment connecting points A and B. Choices (C) is a vertical line where $x = 2$. Choice (E) is a vertical line where $x = 8$. Each of the choices (C) and (E) are vertical lines, but neither would be located between the two given points.

Question 3: Compound Loci

How many points are contained in the intersection of the locus of points 4 units from a given line *l* and 6 units from a given point P located on line *l*?

(A) 0

(B) 1

(C) 2

(D) 4

(E) 6

The correct answer is choice (D). This is a compound loci question in which there are two different sets of criteria for the locus of points. Find the locus of points for each set of criteria and look for the intersection of the loci.

The first step is to draw a diagram of the given objects. This problem involves a line *l* and a point P located on line *l*. The figure below shows the diagram for this situation.

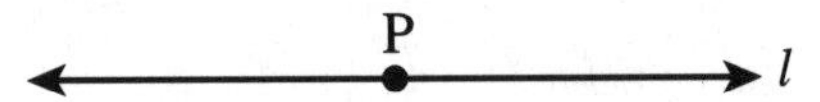

Next, find the locus of points for each condition and draw the locus on the diagram. Since certain measurements were given, try to draw the figures as much to scale as possible. The locus of points 4 units from line *l* is two lines parallel to *l* on either side of *l*. Each of the lines is 4 units from *l*. The locus of points 6 units from point P is a circle with point P as the center and a radius of 6 units. The compound loci are shown as dotted lines in the figure below.

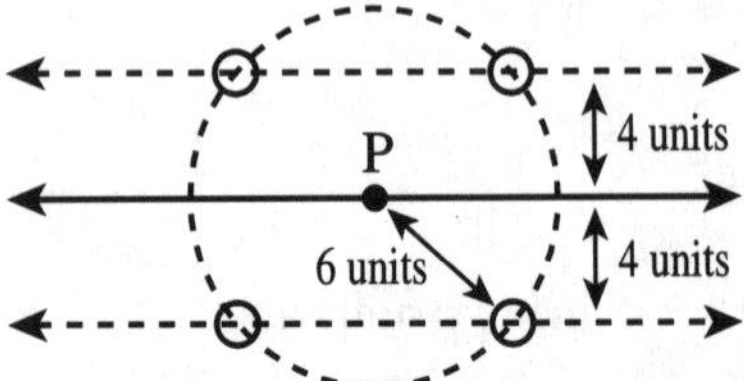

As shown in the above figure, the dotted lines intersect in four places, so there are 4 points in the intersection of the locus of points for each situation.

Question 4: Locus of Points About a Line in the Coordinate Plane

Which best describes the locus of points three units from the line $y = 4$?

(A) $y = 1$

(B) $x = 1$

(C) $y = 1$ and $y = 7$

(D) $x = 1$ and $x = 7$

(E) $y = 4x$

The correct answer is choice (C). The locus of points equidistant from a given line comprises two lines on either side of the given line. Each of these two lines is parallel to the given line. In this problem, each line in the locus of points is 3 units from the given line. The given line $y = 4$ is a horizontal line four units above the x-axis. To find the lines that make up the locus of points, add $3 + 4 = 7$ and subtract $4 - 3 = 1$. Thus, the equations of the lines that make up the locus of points are $y = 7$ and $y = 1$. The given line and the locus of points are graphed in the figure below.

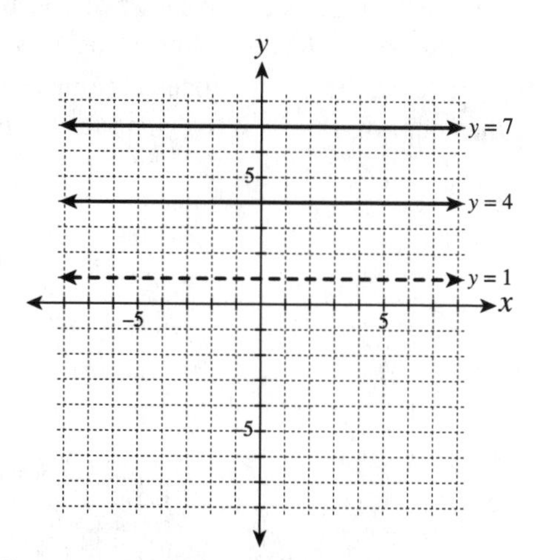

Choice (A) only lists one of the lines in the locus of points. Choice (B) is a vertical line at $x = 1$ and would not be parallel to the given horizontal line. Choice (D) is a vertical line at $x = 1$ and $x = 7$ and would not be parallel to the given line. Choice (E) is a diagonal line that crosses at the origin and has a slope of 4. This also would not be parallel to the given line in the problem and not form the set of points three units away.

Question 5: Equation of a Circle

Which of the following equations best describes the locus of points 4 units from the origin?

(A) $x^2 + y^2 = 2$

(B) $x^2 + y^2 = 4$

(C) $y = x^2 + 16$

(D) $x = y^2 + 4$

(E) $x^2 + y^2 = 16$

The correct answer is choice (E). The set of points equidistant from a single point is a circle that surrounds the point. The radius of the circle is the distance the locus is from the given point. The equation of a circle with the center at the origin is $x^2 + y^2 = r^2$, where r represents the radius of the circle. The equation of a circle with the center at (h, k) is $(x - h)^2 + (y - k)^2 = r^2$, where r represents the radius of the circle.

When locating the locus of points equidistant from a single point in the coordinate plane, use the given point as the center of the circle. In this question, the center is the origin and the radius of the circle is 4. Using the first equation in the above paragraph, the locus of points 4 units from the origin would have the equation $x^2 + y^2 = 16$. This locus of points is shown in the figure below.

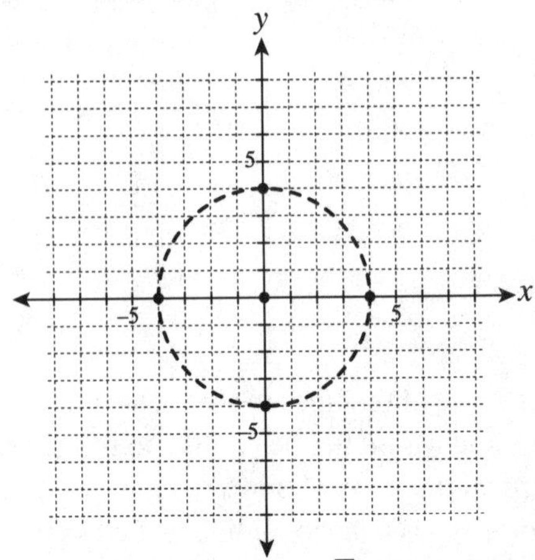

Choice (A) describes the locus of points $\sqrt{2}$ units from the origin. Choice (B) is the equation of a circle that is 2 units from the origin. You may have selected this answer choice if you forgot to evaluate r^2 and substituted $r =$

4 into the equation. Choice (C) is the equation of a parabola that opens up, not the equation of a circle. Choice (D) is the equation of a parabola that opens to the right and does not represent the equation of a circle. Each of the equations in choices (C) and (D) do not include both x^2 and y^2.

CHAPTER QUIZ

1. Which of the following best describes the locus of points 8 units from a given point A?

   (A) A circle with a radius of 8 units that contains point A on the circle.

   (B) A circle with a radius of 8 units where A is the center of the circle.

   (C) A circle with a radius of 16 units that contains point A on the circle.

   (D) A circle with a radius of 16 units where A is the center of the circle.

   (E) A circle with a radius of 64 units where A is the center of the circle.

2. What is the equation of the locus of points equidistant from C(−2, 5) and D(−2, 9)?

   (A) $y = 7$

   (B) $x = 7$

   (C) $y = -2$

   (D) $x = -2$

   (E) $y = 14$

3. Which of the following best illustrates the locus of points about a given line *t*?

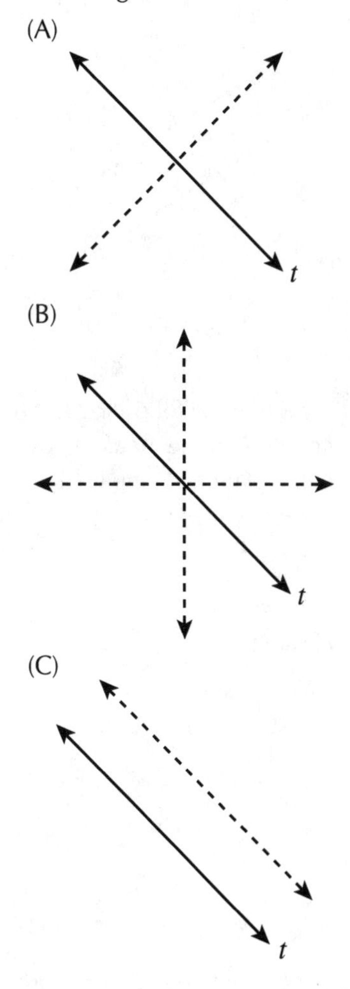

(D)

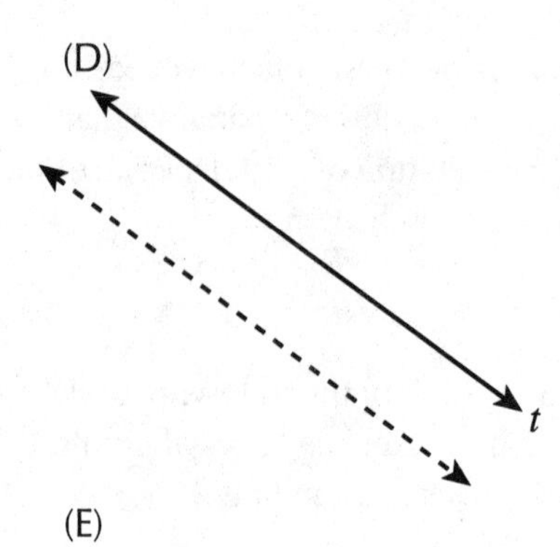

(E)

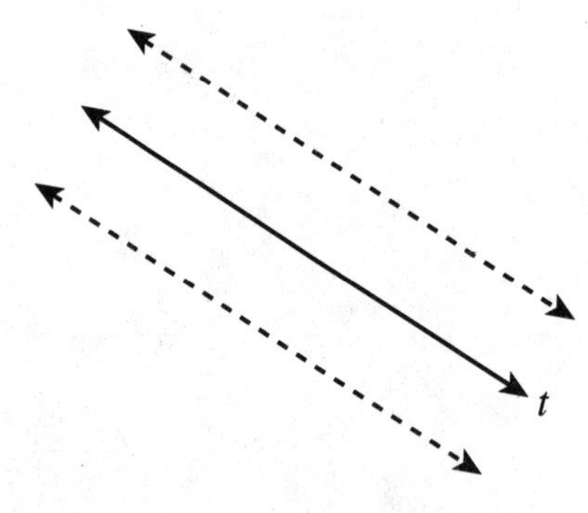

4. Which of the following is the equation of the locus of points equidistant from the lines $x = -1$ and $x = -11$?

   (A) $x = 6$

   (B) $x = -12$

   (C) $x = -6$

   (D) $y = -6x$

   (E) $y = -6$

5. Which of the following best represents the locus of points equidistant from both the *x*-axis and the *y*-axis?

   (A) $y = x$

   (B) $y = -x$

   (C) $y = 0$ and $x = 0$

   (D) $y = x$ and $y = -x$

   (E) $y = 0$

6. Which of the following best describes the locus of points equidistant from acute angle $\angle ABC$?

   (A) One point in the interior of angle $\angle ABC$

   (B) Three points on angle $\angle ABC$

   (C) The bisector of angle $\angle ABC$

   (D) The perpendicular bisector of ray BC

   (E) The perpendicular bisectors of ray BC and ray BA

7. How many points are contained in the intersection of the locus of points 7 units from a given line *l* and 3 unjts from a given point P located on line *l*?

   (A) 0

   (B) 1

   (C) 2

   (D) 3

   (E) 4

8. Which of the following does not state the coordinates of the locus of points 5 units from the origin and 4 units from the x-axis?
   (A) (3, 4)
   (B) (−3, 4)
   (C) (4, −3)
   (D) (−3, −4)
   (E) (3, −4)

9. What is the equation of the locus of points 9 units from the point (−1, 8)?
   (A) $(x + 1)^2 + (y - 8)^2 = 9$
   (B) $(x + 1)^2 + (y - 8)^2 = 81$
   (C) $(x - 1)^2 + (y - 8)^2 = 9$
   (D) $(x - 1)^2 + (y + 8)^2 = 9$
   (E) $(x - 1)^2 + (y + 8)^2 = 81$

10. The locus of points about a point in the coordinate plane is represented by the equation $(x - 4)^2 + (y + 5)^2 = 49$. What is the center and radius of this circle?
   (A) Center at (−4, 5) and radius is 49 units
   (B) Center at (−4, 5) and radius is 7 units
   (C) Center at (4, −5) and radius is 49 units
   (D) Center at (4, −5) and radius is 7 units
   (E) Center at (4, −5) and radius is 24.5 units

ANSWER EXPLANATIONS

1. B

The locus of points about a point is a circle that surrounds the point. The given point is the center of the circle and the radius is the number of units each point is equidistant from the given point. In this case, the center of the circle is A and the radius is 8 units, which is choice (B).

Choice (A) has the correct radius but incorrectly states that point A should be on the circle. Choice (C) incorrectly has a radius of 16 units and states that point A is located on the circle. Choice (D) has an incorrect radius of 16 units. Choice (E) may have been chosen if the value of the radius was incorrectly squared, as it is in the equation of the circle.

2. A

The locus of points equidistant from two given points is a line between them. This line would be the perpendicular bisector of the line segment that would connect them. Point C and point D are located in quadrant III. The line that represents the locus of points is a horizontal line between them and is represented by an equation of the form $y = k$. Since point C is five units above the x-axis and point D is nine units above the x-axis, the equation of the line between them is $y = 7$. This is shown in the figure below.

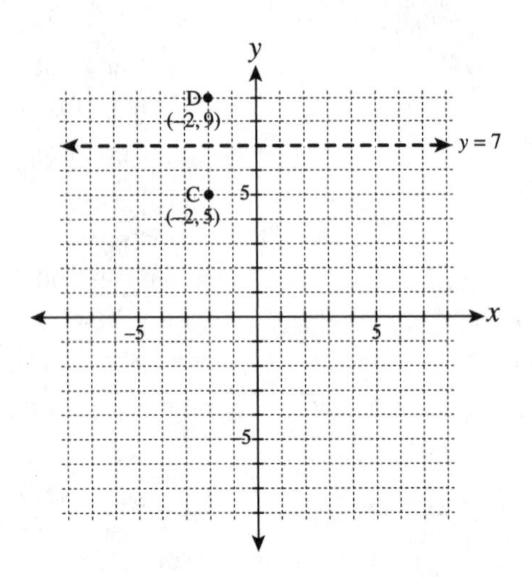

Choice (B) is a vertical line seven units to the right of the y-axis. Choice (C) is a horizontal line two units below the x-axis. Choice (D) is a vertical line two units to the left of the y-axis. This line would pass through points C and D, but would not form the set of points equidistant from the points. Choice (E) is a horizontal line 14 units above the x-axis and may have been chosen if the y-coordinates for each point were incorrectly added together.

3. E

The locus of points equidistant from a given line is two lines parallel to the given line. Each line is the same distance from the given line. Since the locus of points forms two parallel lines, these lines will not cross the given line. The only answer choice with two lines parallel to the given line is choice (E).

Choice (A) shows one line that intersects the given line. Choice (B) shows the locus as two lines that intersect the given line, and the locus is two lines that are parallel to the given line. Choices (C) and (D) show lines that are parallel to the given line, but each answer choice only has one line, and there should be two.

4. C

The line $x = -1$ is a vertical line parallel to the y-axis that is located one unit to the left of the y-axis. The line $x = -11$ is a vertical line parallel to the y-axis that is located 11 units to the left of the y-axis. The locus of points equidistant from two parallel lines is a parallel line located between the two lines that is the same distance from each line. These two lines are 10 units apart, so the locus of points will be five units from each line, and between the two lines. Therefore, the locus of points will be located at $x = -6$.

Choice (A) is a vertical line where $x = 6$. Choice (B) may be the result of adding the values for the given lines. Choice (D) is a diagonal line with a slope of -6 and a y-intercept of 0. This line would not be equidistant from the two given lines. Choice (E) is a horizontal line at $y = -6$, and would pass through the given lines but not be parallel to them.

5. D

The x- and y-axis are two intersecting lines. The locus of points equidistant from two intersecting lines is two other intersecting lines. The lines that form the locus will bisect the vertical angles formed by the original lines. Since the x- and y-axis are perpendicular lines, the locus of points will be at 45-degree angles to the axes. These two diagonal lines will have a slope of 1 and -1 and intersect the y-axis at the origin. Thus, the equations of these intersecting lines are $y = x$ and $y = -x$.

Choice (A) correctly lists one of the two equations that form the locus of points, but is missing the other equation. Choice (B) also lists one of the correct equations for the locus, but both need to be named together in the answer choice. Choice (C) lists the equations for the x- and y-axes themselves, not the locus of points equidistant from both lines at the same time. Choice (E) is the equation for the x-axis.

6. C

To find the locus of points equidistant from the rays that form an angle, consider the rays as part of two intersecting lines. Keep in mind that the angle in the question is acute, so the measure of the angle is between 0 and 90 degrees. Therefore, the locus of points will be a ray with the same vertex as the angle that bisects the angle. An example of an acute angle $\angle ABC$ with the locus of points drawn as a dotted line is shown in the figure below.

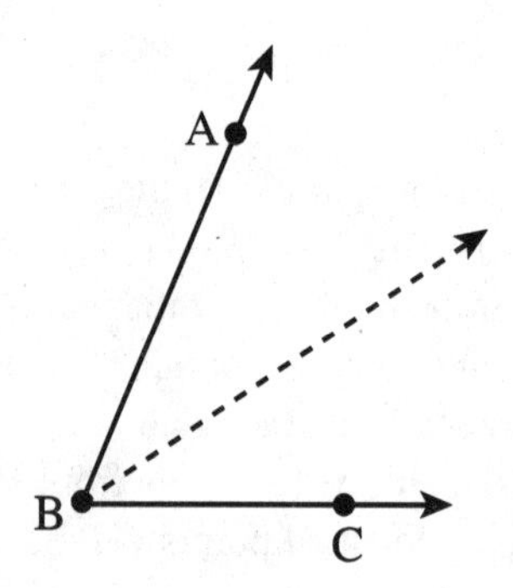

7. A

This question involves compound loci. In other words, it describes the locus of points that satisfy two or more conditions at the same time. The locus of points equidistant from a given line is two lines parallel to the given line on either side of the line. The locus of points 7 units from a given line l are two lines, one on either side of l. Each line is 7 units from l and is parallel to l. The locus of points equidistant from a given point is a circle that surrounds that point. The locus of points 3 units from a point P located on line l is a circle with radius of 3 units where P is the center. A picture of this situation is shown below. The locus of points is drawn as dotted lines in the figure.

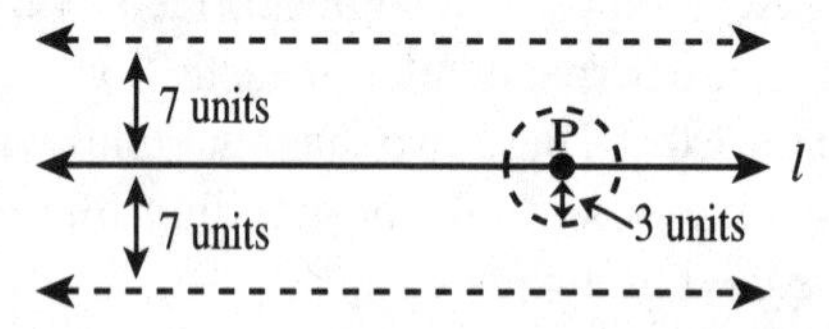

Notice that the locus of points does not intersect, therefore, the intersection of the locus is zero points.

8. C

This question deals with compound loci. There are two conditions to be met. The first condition is the set of points 5 units from the origin. This is a circle with radius of 5 units with center at the origin. The second condition is the set of points 4 units from the x-axis. This locus is two horizontal lines parallel to the x-axis, with one 4 units above the x-axis and one 4 units below the x-axis. These are the lines $y = 4$ and $y = -4$. Each of the compound loci is graphed in the figure below as dotted lines.

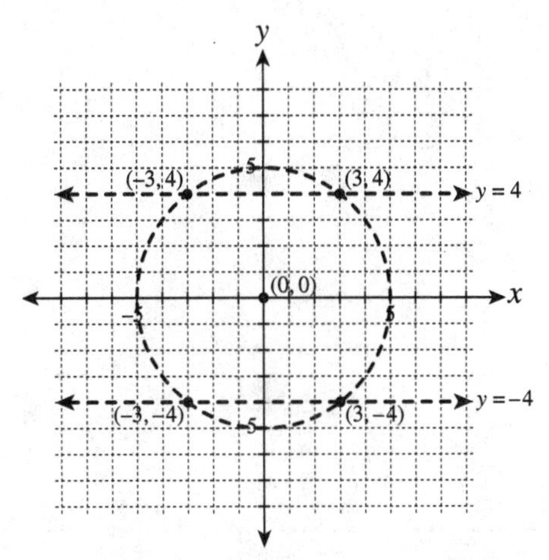

The loci intersect at four locations: (3, 4), (−3, 4), (−3, −4), and (3, −4). Therefore, the location where the loci do not intersect is choice (C), (4, −3).

9. B

The set of points equidistant from a single point is a circle that surrounds the points. The radius of the circle is the distance the locus is from the given point. The equation of a circle with the center at the origin is $x^2 + y^2 = r^2$, where r represents the radius of the circle. The equation of a circle with the center at (h, k) is $(x - h)^2 + (y - k)^2 = r^2$, where r represents the radius of the circle.

When locating the locus of points equidistant from a single point in the coordinate plane, use the given point as the center of the circle. In this question, the center is the point (−1, 8) and the radius of the circle is 9. Using the second equation in the above paragraph, the locus of points 9 units from the point (−1, 8) would have the equation $(x + 1)^2 + (y - 8)^2 = 81$. This locus of points is shown in the figure on the following page.

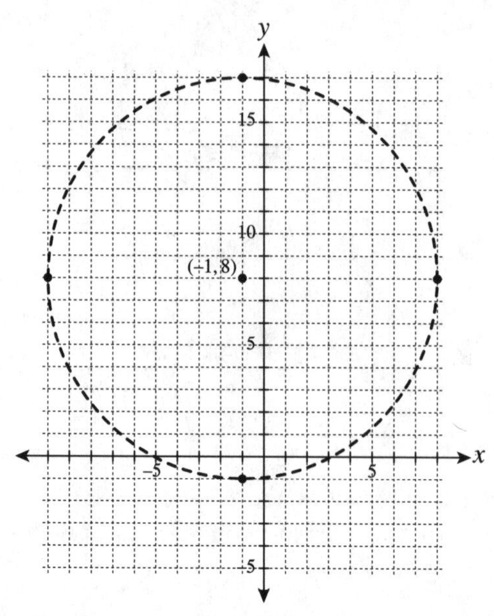

Choice (A) describes the locus of points 3 units from the point (−1, 8). Don't forget to use r^2 and not just r in the equation of a circle. Choice (C) is the equation of a circle that is 3 units from the point (1, 8). Choice (D) is the locus of points 3 units from the point (1, −8). Choice (E) describes the locus of points 9 units from the point (1, −8).

10. D

The equation of a circle with the center at the origin is $x^2 + y^2 = r^2$, where r represents the radius of the circle. The equation of a circle with the center at (h, k) is $(x - h)^2 + (y - k)^2 = r^2$, where r represents the radius of the circle. This equation is $(x - 4)^2 + (y + 5)^2 = 49$. Therefore, this is the equation of a circle with center at the point (4, −5) and a radius of 7.

Choice (A) would be the equation $(x + 4)^2 + (y - 5)^2 = 2{,}401$ and may have been chosen if the signs were switched on the center point and the value of r^2 was mistaken for the value of r. Choice (B) would be the equation $(x + 4)^2 + (y - 5)^2 = 49$ and may have been selected if the signs of the center point were switched. Choice (C) would be the equation $(x - 4)^2 + (y + 5)^2 = 2{,}401$ and may have been selected if the value of r^2 was mistaken for the value of r. Choice (E) is the equation $(x - 4)^2 + (y + 5)^2 = 600.25$ and may have been selected if r^2 was divided by two, instead of taking the square root to find r.

CHAPTER 12

Right Triangle Geometry

WHY IS THE RIGHT TRIANGLE SO IMPORTANT?

Of all of the figures in geometry, the right triangle plays a critical role. The right triangle is used for measurement, both in the real world and throughout geometry. Right triangles form the basis of many real-world applications, such as finding the height of objects by using similarity or trigonometry. The right triangle is also used to find the shortest distance between two points on a coordinate plane. Many concepts throughout this book are related to right triangles.

CONCEPTS TO HELP YOU

This section will identify the key parts of a right triangle and explain one of the most useful theorems in all of mathematics—the Pythagorean Theorem. Two special right triangles will be introduced, as well as trigonometry and the mean proportional with right triangles.

The Parts of a Right Triangle

Right triangles are triangles that contain one right angle and two acute angles. Angle classification was explained in Chapter 1. The side of the right triangle across from the right angle is called the *hypotenuse.* The other two sides are called the *legs*.

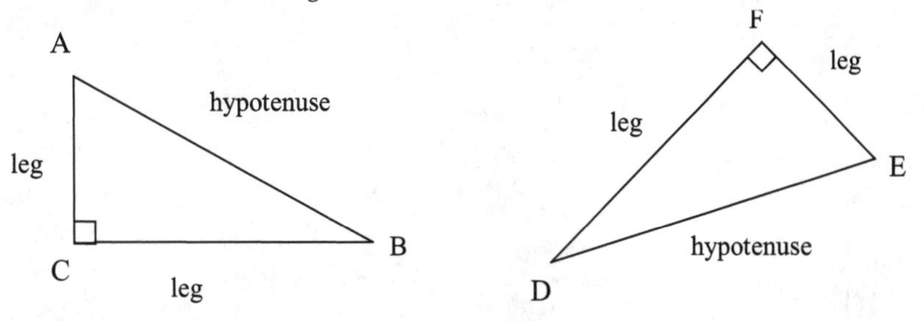

In the figures above, sides AB and DE are the hypotenuses. The hypotenuse is always opposite to the right angle, and therefore it is always the longest side. Leg AC is opposite to angle ∠B and adjacent to angle ∠A. Leg FE is opposite to angle ∠D and adjacent to angle ∠E.

The Pythagorean Theorem

One of the most useful theorems in all of mathematics is the *Pythagorean Theorem*, which describes the relationship between the hypotenuse and the legs of a right triangle.

THE PYTHAGOREAN THEOREM

For every right triangle, the sum of the squares of the legs is equal to the square of the hypotenuse. Traditionally, the hypotenuse of a right triangle is denoted by the variable *c* and the legs as the variables *a* and *b*. Using algebra, the Pythagorean Theorem is: $a^2 + b^2 = c^2$

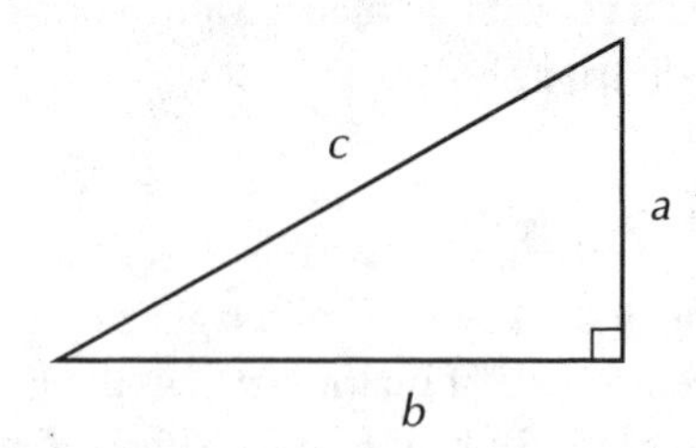

For example, to find the length of the missing side AB in the triangle below, use the Pythagorean Theorem. Let x represent the length of AB, which is the hypotenuse of triangle ABC. Solve the equation $x^2 = 6^2 + 4^2$, or

$x = \sqrt{36 + 16} = \sqrt{52}$.

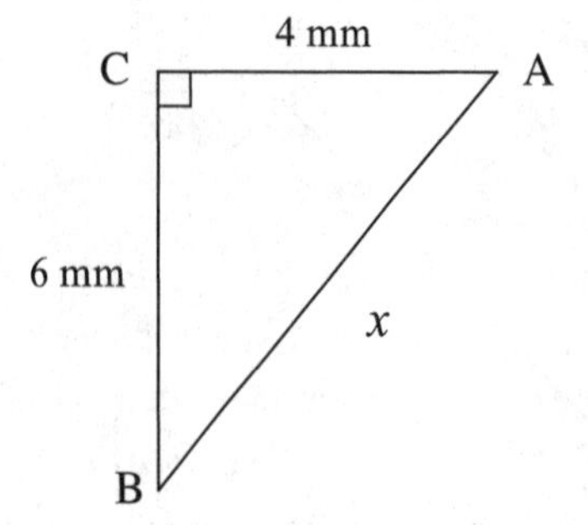

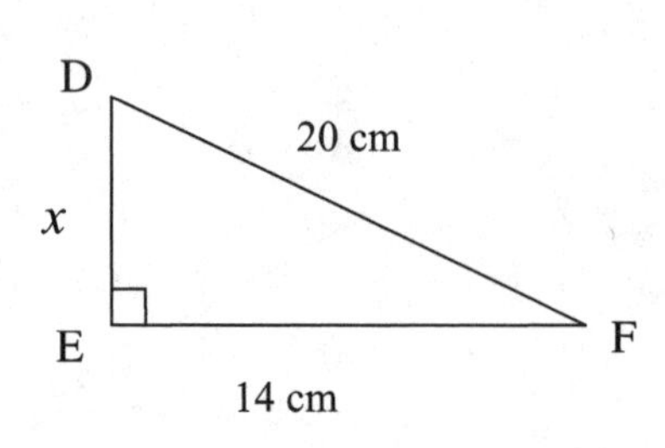

To find the length of a leg, you also use the Pythagorean Theorem. In triangle DEF above, the missing measurement is leg DE. Let x represent this length and set up the equation $20^2 = 14^2 + x^2$. By solving the equation for x, leg DE is $x = \sqrt{20^2 - 14^2}$. This simplifies to $x = \sqrt{400 - 196} = \sqrt{204} = 2\sqrt{51}$.

The converse of the theorem is also true.

THE CONVERSE OF THE PYTHAGOREAN THEOREM

If, in a triangle, the sum of the squares of the two shortest sides is equal to the square of the longest side, then the triangle is a right triangle.

For example, if you are given triangle side lengths of 10, 24, and 26, you can verify that it is a right triangle because $26^2 = 10^2 + 24^2$, which is $676 = 100 + 576$, or $676 = 676$.

Special Right Triangles

Some right triangles have side measurements that are whole numbers. These combinations of side measurements are called *Pythagorean triples.* Some examples are the 3-4-5 right triangle, the 5-12-13 right triangle and the 8-15-17 right triangle. These triangle measures appear regularly in all kinds of testing situations. Similar triangles were explained in Chapter 2. Because of similarity, any triangles whose sides are proportional to these triples are also Pythagorean triples. For example, the 9-12-15 right triangle is a Pythagorean triple because $3 \cdot 3 = 9$, $4 \cdot 3 = 12$, and $5 \cdot 3 = 15$.

Another special triangle is the *isosceles right triangle*, also known as the 45-45-90 degree triangle.

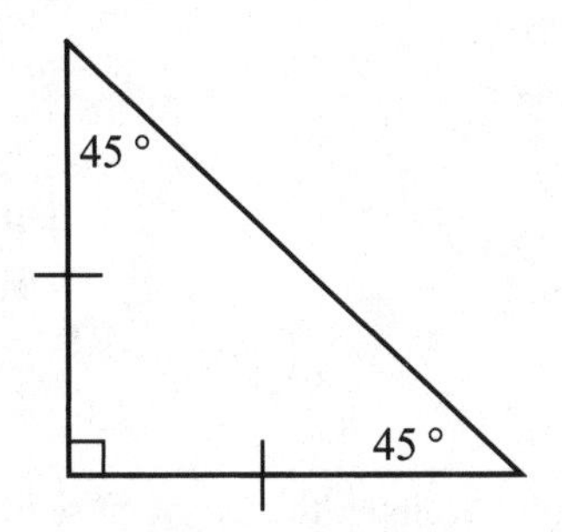

Because the legs of an isosceles right triangle are equal, you can use the Pythagorean Theorem to show a helpful fact about this triangle. Using the theorem, let x represent the length of the legs.

Write the theorem: $a^2 + b^2 = c^2$

Substitute in the variable x: $x^2 + x^2 = c^2$

Simplify to get: $2x^2 = c^2$

Take the square root of both sides: $c = x\sqrt{2}$

So, for an isosceles right triangle, the hypotenuse is $\sqrt{2}$ times the length of the legs.

The other common special right triangle is the 30-60-90 degree right triangle. To find the special relationship among the sides, take an equilateral triangle and draw a line of symmetry. Symmetry was explained in Chapter 10. Two congruent triangles are formed, as shown below:

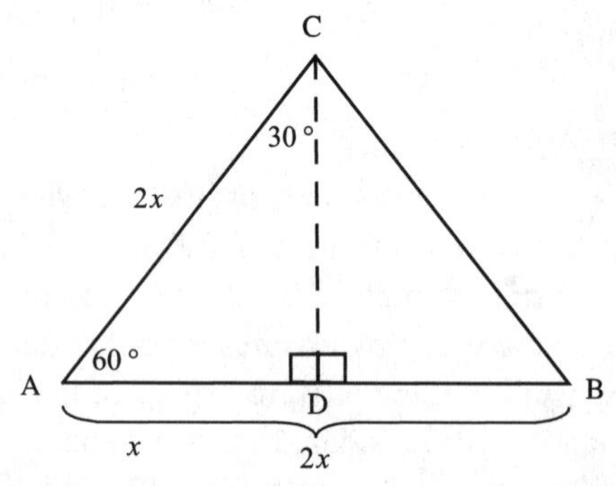

Notice that two congruent right triangles are formed. Segment AD is one-half the length of side AB because of symmetry. Let x represent the length of side AD, and therefore $2x$ represents the length of side AB, and also side AC, because the large triangle is equilateral. Use the Pythagorean Theorem to find the length of BD:

Write the theorem: $a^2 + b^2 = c^2$

Substitute in the variable x: $x^2 + BD^2 = (2x)^2$

Simplify to get: $x^2 + BD^2 = 4x^2$

Subtract x^2 from both sides: $x^2 + BD^2 - x^2 = 4x^2 - x^2$

Simplify: $BD^2 = 3x^2$

Take the square root of each side: $BD = x\sqrt{3}$

So, for a 30-60-90 degree right triangle, the hypotenuse is twice the length of the shortest leg, and the longer leg is $\sqrt{3}$ times the length of the shortest leg.

Trigonometric Ratios

If you compare the side lengths of right triangles, you find that according to the measure of the acute angles, the sides form ratios that are named the sine, cosine, and tangent of the acute angle. The *sine* function is defined as the ratio of the length of the side that is opposite the angle to the length of the hypotenuse. The *cosine* function is defined as the ratio of the length of the side that is adjacent to the angle, to the length of the hypotenuse. The *tangent* function is defined as the ratio of the length of the side that is opposite the angle to the length of the side adjacent to the angle. These ratios are defined for any value of an acute angle, and they remain constant, because of similarity in right triangles.

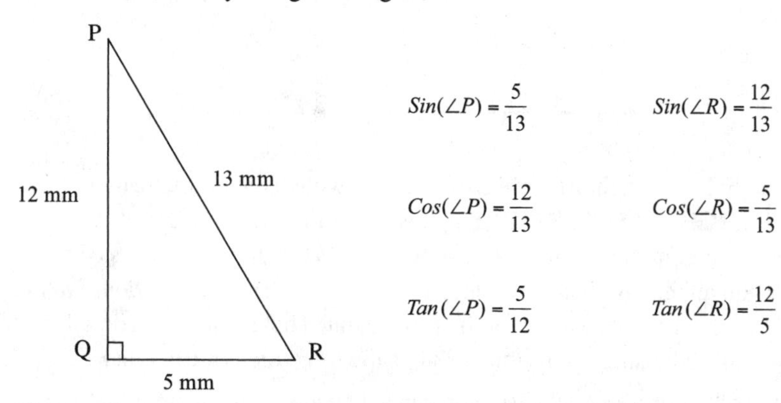

These ratios are expressed as decimal equivalents on a calculator and are found by using the SIN, COS and TAN keys. Likewise, if you want to know the measure of an angle with a specific sine, cosine or tangent ratio, you use the SIN^{-1}, COS^{-1} or TAN^{-1} keys.

THE TRIGONOMETRIC RATIOS

Sine of angle θ is $\frac{opposite}{hypotenuse}$

Cosine of angle θ is $\frac{adjacent}{hypotenuse}$

Tangent of angle θ is $\frac{opposite}{adjacent}$

These ratios can be remembered by the familiar mnemonic SOH-CAH-TOA.

The Mean Proportional

An *altitude* to a triangle is the line segment that has one endpoint at a vertex of the triangle and is perpendicular to the opposite side. When an altitude is drawn from the right angle of a right triangle to the hypotenuse, two smaller right triangles are formed. Because the altitude is perpendicular, the two smaller triangles are also right triangles. Each of these three triangles is similar to one another, and therefore the corresponding sides are in proportion. Study this figure:

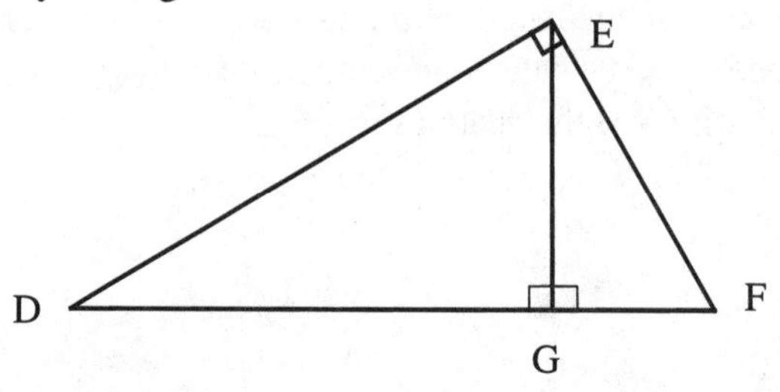

Altitude EG from right triangle DEF creates two smaller similar right triangles DGE and EGF. Recall that in a proportion, the product of the means is equal to the product of the extremes. When the two means of a proportion are equal, then either mean is called the *mean proportional* of the extremes, the other two terms of the proportion. The mean proportional is this altitude. The altitude, segment EG, is the longer leg of triangle EGF and the shorter leg of triangle DGE. You can set up a proportion with the legs of the two smaller triangles as $\frac{long}{short} = \frac{long}{short}$. Substitute in the segments in the figure, and you have $\frac{DG}{EG} = \frac{EG}{GF}$. By cross multiplication, $(EG)^2 = (DG)(GF)$.

THE MEAN PROPORTIONAL IN A RIGHT TRIANGLE

In a right triangle, the square of the length of the altitude from the right angle is equal to the product of the segments of the hypotenuse.

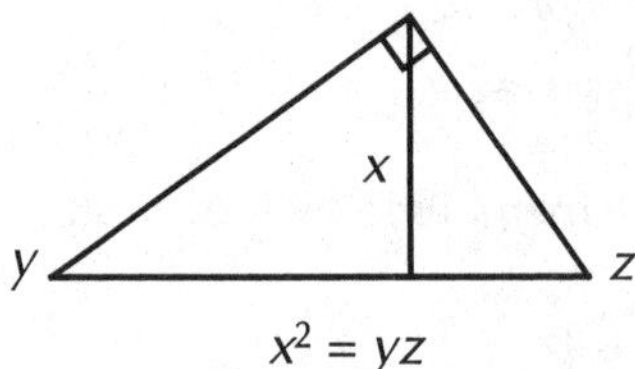

There are other proportions that can be set up with these three similar right triangles that can be used in problem solving. For example, triangle EDG is similar to triangle EFD below.

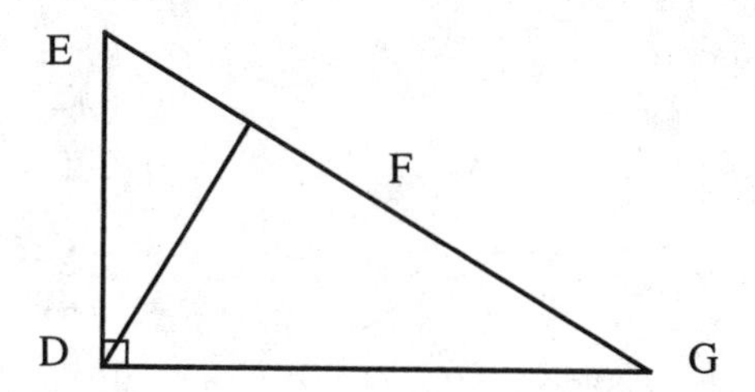

Because of this similarity, you can set up the proportion $\frac{\text{hypotenuse}}{\text{shorter leg}} = \frac{\text{hypotenuse}}{\text{shorter leg}}$, or $\frac{EG}{ED} = \frac{ED}{EF}$.

By cross multiplication, $(EG)(EF) = (ED)^2$.

STEPS YOU NEED TO REMEMBER

This section will walk you through different applications that deal with right triangles. The various steps involved in right triangle geometry are illustrated. After reviewing this material, you will be on your way to mastering this topic.

Using the Pythagorean Theorem

When you are given a right triangle with two side lengths and need to find the missing side, you use the Pythagorean Theorem, $a^2 + b^2 = c^2$, to find the measure. First, identify the hypotenuse, c, and the legs, a and b. Write the formula and substitute in the given information. Then, solve for the missing sid e.

For example, find the length of HI in triangle GHI below:

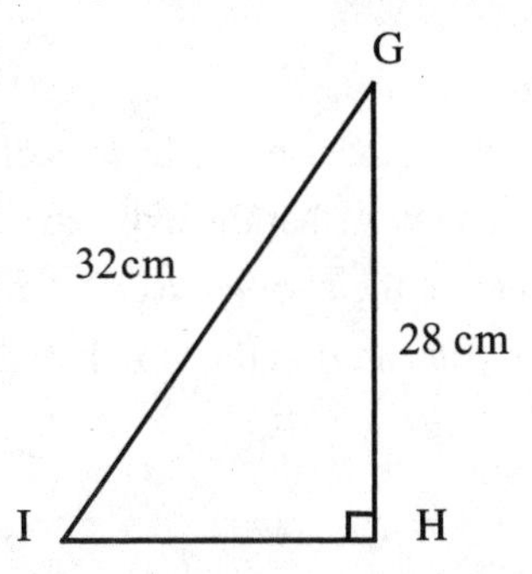

Side GI, of length 32 cm, is the hypotenuse and GH, of length 28 cm, is one of the legs. HI is the missing leg length.

Write the Pythagorean Theorem:	$a^2 + b^2 = c^2$
Substitute into the theorem:	$28^2 + b^2 = 32^2$
Simplify:	$784 + b^2 = 1024$
Subtract 784 from both sides:	$b^2 = 240$
Take the square root of both sides:	$\sqrt{b^2} = \sqrt{240}$
	$b = 15.49$

So side HI is 15.49 cm, rounded to the nearest hundredth.

Another use of the Pythagorean Theorem is to determine if given triangle lengths are sides of a right triangle. To do this, test the sides by substituting into the Pythagorean Theorem. Remember that the longest side is always the hypotenuse. For example, to test if a triangle with sides 9, 11, and 15 is a right triangle, test if $9^2 + 11^2 = 15^2$. Simplify to get $81 + 121 = 225$, add on the left-hand side and you can see that $202 \neq 225$. Therefore, this triangle is not a right triangle.

To test if a triangle with sides 3, 4, and 5 is a right triangle, test if $3^2 + 4^2 = 5^2$. Simplify to get $9 + 16 = 25$, and because $25 = 25$, these side lengths do form a right triangle.

Working with the Isosceles Right Triangle

If you remember the relationship of the sides of a right isosceles triangle, you can save time in solving problems, by avoiding the need to use the Pythagorean Theorem. In an isosceles right triangle, the legs are congruent and the hypotenuse is $\sqrt{2}$ times the length of the legs. These are called 45-45-90 degree right triangles. If you encounter a problem that has a right triangle with one acute angle of measure 45°, then you have this special relationship.

For example, if you are given that one of the legs of an isosceles right triangle is 16 cm, you then know that the hypotenuse is $16\sqrt{2}$ cm long. Likewise, if you are told that the hypotenuse of a different isosceles right triangle is 14.14 inches, you can find the length of a leg by dividing 14.14 by $\sqrt{2} \approx 10$ inches.

If you encounter a right triangle problem that doesn't seem to give enough information, this is your opportunity to discover if the triangle is a right isosceles triangle. If the two legs are congruent, then use the above relationship to find the missing sides.

Problem Solving with the 30-60-90 Degree Triangle

Another special triangle is the 30-60-90 degree right triangle. If you encounter a problem with a right triangle, and you are told that one of the angles is either 30° or 60°, then remember the relationship of the sides: The hypotenuse is twice the length of the shortest leg, the side opposite to the 30° angle, and the longer leg, opposite to the 60° angle, is $\sqrt{3}$ times the length of the shortest leg. Again, remembering this relationship can be a time saver; you do not have to use the Pythagorean Theorem to solve.

For example, if you are given the right triangle below, you can find the missing two side lengths if you identify it as a 30-60-90 right triangle. Because the shortest side, the side opposite the 30° angle, is 8 cm long, the hypotenuse is $8 \bullet 2 = 16$ cm and the other leg is $8\sqrt{3}$ inches.

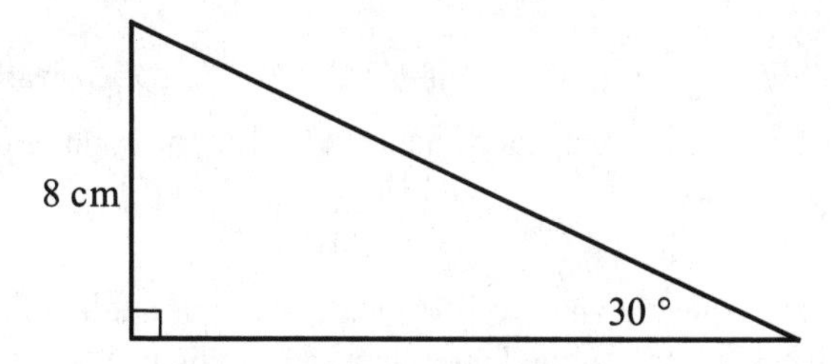

If you encounter a right triangle problem that doesn't seem to give enough information, this is your opportunity to discover if the triangle is the 30-60-90 degree triangle and use the above relationship to find the missing sides.

Working with Trigonometric Ratios

There are two situations where trigonometry is used to solve right triangle problems:

1. If you are given a right triangle with an acute angle measure, one side length, and are asked to find another side of the triangle
2. If you are asked to find the measure of an acute angle and are given two side measures

To solve these problems, identify the given sides as either the hypotenuse, or the sides opposite or adjacent to the given angle. This will determine what trigonometric function to use. Recall that sine compares the opposite to the hypotenuse, cosine compares the adjacent to the hypotenuse, and tangent compares the opposite to the adjacent.

In the triangle below you can find the measure of a missing side of the triangle, such as side XY, by using a trigonometric ratio.

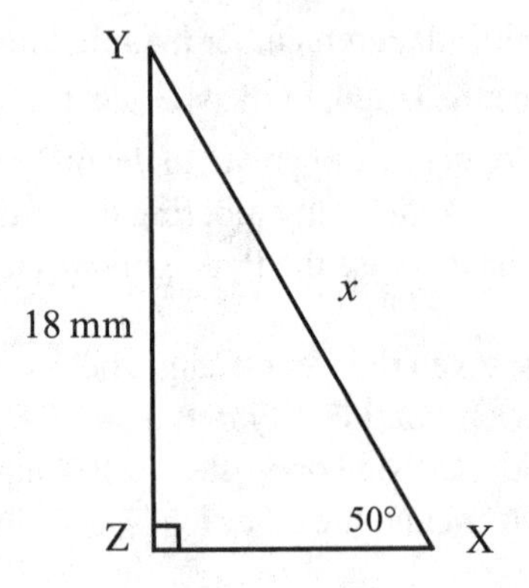

Because you know the measure of angle YXZ = 50°, and the measure of the side opposite is 18 mm, use the sine function to solve for hypotenuse XY. The equation is $Sin(50°) = \frac{18}{x}$, or $x = \frac{18}{Sin(50°)}$, or $x \approx 23.50$.

In the triangle below, if you want to find the measure of angle MLN, given the lengths of the side adjacent and the hypotenuse, you can find this measure by the equation $Cos(\theta) = \frac{8}{12}$, or $\theta = Cos^{-1}\left(\frac{8}{12}\right) \approx 48.19°$

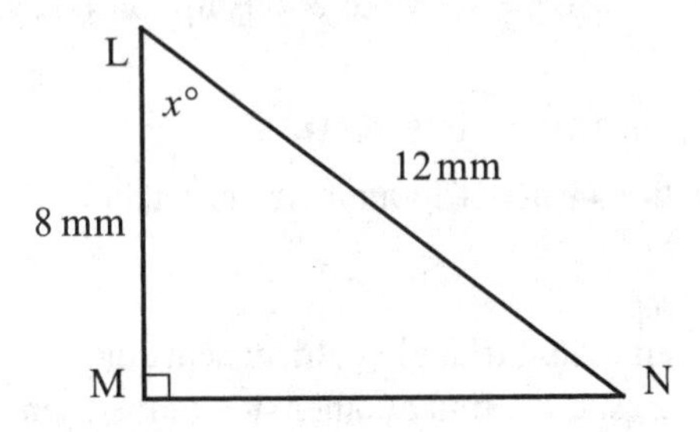

Common real-world applications for trigonometry are problems dealing with angles of elevation or depression. Angles of elevation are used when a surveyor, for example, sites the top of a tall structure, as shown below:

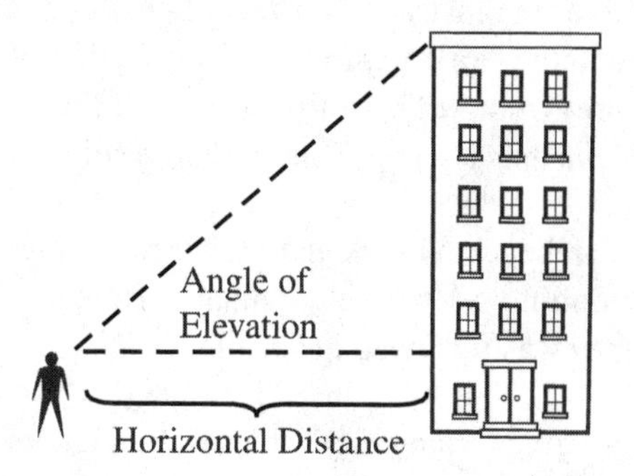

If you are told the angle of elevation and you know the horizontal distance, you can use the tangent function to find the height of the building.

Angles of depression occur when the surveyor is up at the height and sites something down below at some horizontal distance away, as shown here:

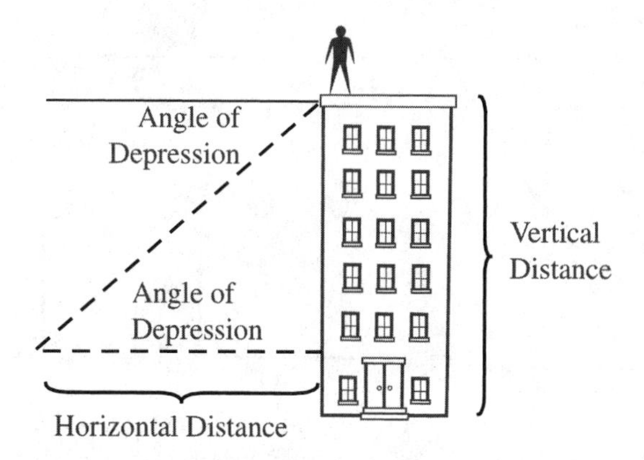

Note the parallel lines shown are cut by the transversal, which is the hypotenuse of the right triangle. Parallel lines and congruent angles were reviewed in Chapter 1. The angle of depression is shown above. If you are given the angle measure and one other side measure, you would use one of the trigonometric functions to find a missing measure, as shown in the examples earlier in this section.

Using the Mean Proportional

When an altitude is drawn to the hypotenuse of a right triangle, you use the mean proportional to find missing measures. Remember the special proportions described in the Concepts to Help You section of this chapter and that each of the three triangles is similar to each of the others. Use the facts outlined in the Concepts to Help You section to solve problems.

Another approach is that you can break apart the three triangles and determine the corresponding sides to set up many different proportional relationships. Similarity was reviewed in Chapter 2.

For example, if you are given triangle JKL below, break it up into one big triangle JKL, one medium triangle KML and one small triangle JMK.

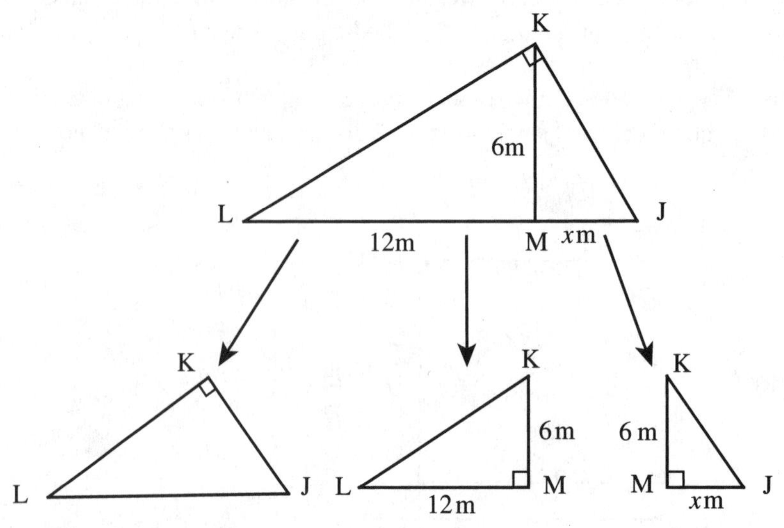

You can determine the corresponding sides by identifying the hypotenuses, the shorter legs, and the longer legs of each triangle. For each of the three triangles, the hypotenuses correspond, that is JL in triangle JKL, JK in triangle JMK, and KL in triangle KML. Likewise, the shortest sides correspond, that is JK in triangle JKL, JM in triangle JMK and KM in triangle KML. The longer legs are KL in triangle JKL, MK in triangle JMK and ML in triangle KML.

An example of using the mean proportional would be to find the length of segment MJ in the figure above. The mean proportion which is altitude KM squared is equal to the product of the segments of the hypotenuse, LM and MJ.

Substitute the given measures: $6^2 = 12x$
Square the left-hand side to get: $36 = 12x$
Divide both sides by 12: $\frac{36}{12} = \frac{12x}{12}$
Simplify: $3 = x$

The measure of the shorter segment of the hypotenuse, MJ, is equal to 3m.

STEP-BY-STEP ILLUSTRATION OF THE FIVE MOST COMMON QUESTION TYPES

Question 1: The Pythagorean Theorem Applied to a Real-World Problem

A 15-foot ladder rests against the side of a building. The base of the ladder is 9 feet from the base of the building. How far up the building does the ladder reach?

(A) 3.46 feet

(B) 6 feet

(C) 6.93 feet

(D) 12 feet

(E) 17.49 feet

A ladder resting against a building is the hypotenuse of a right triangle, and the distance from the base of the building to the base of the ladder is one of the legs.

Use the Pythagorean Theorem: $a^2 + b^2 = c^2$
Substitute into the theorem: $9^2 + b^2 = 15^2$
Simplify: $81 + b^2 = 225$
Subtract 81 from both sides: $b^2 = 144$
Take the square root of both sides: $\sqrt{b^2} = \sqrt{144}$

$b = 12$

The ladder reaches 12 feet up the side of the building. **The correct answer choice is (D).** If you chose answer (A), you multiplied 15 and 9 by two instead of squaring them. If you chose (B), you multiplied 15 and 9 by two, and then divided this difference by two. Answer choice (C) would be arrived at by taking the square root of [(15 • 2) + (9 • 2)]. Choice (E) would be the length of a hypotenuse if the legs were 15 and 9.

Question 2: Special Right Triangles

In triangle ABC below, what is the length of CB?

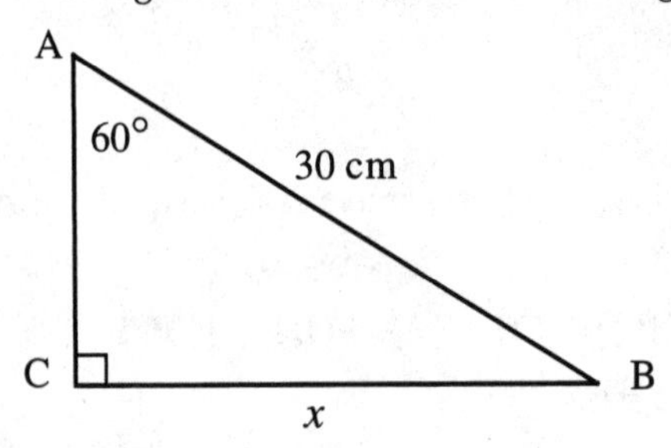

(A) 15 cm

(B) $15\sqrt{2}$ cm

(C) $15\sqrt{3}$ cm

(D) $30\sqrt{3}$ cm

(E) 60 cm

This triangle is a 30-60-90 degree right triangle. In these special triangles, the side opposite the 30° angle is one-half the length of the hypotenuse, or 15 cm. The side asked for, side CB, is the side opposite the 60° angle, so it is $15\sqrt{3}$ cm in length. **The correct answer choice is (C).** Choice (A) is the length of AC. If your answer was (B), you may have confused this triangle with a right isosceles triangle. In choice (D), you used the hypotenuse instead of the shortest leg. Choice (E) is twice the hypotenuse.

Question 3: Finding a Missing Side Using Trigonometry

In triangle RST, what is the length of ST?

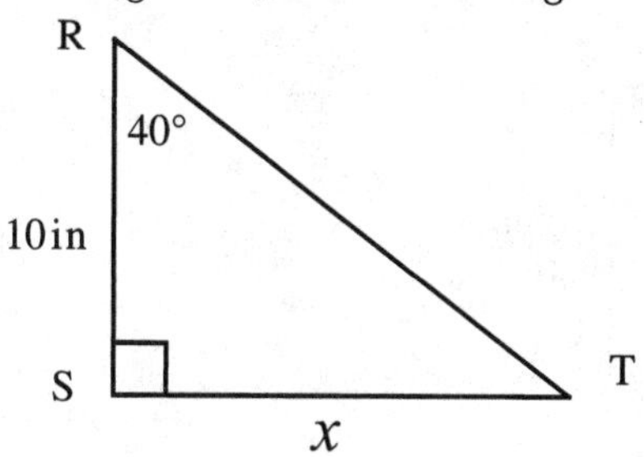

(A) 10 inches

(B) 11.92 inches

(C) 15.56 inches

(D) 38.73 inches

(E) 83.90 inches

In this triangle, you are given an angle measure and the length of the side opposite this angle. Use the trigonometric ratio of tangent, because you need to find the length of the adjacent side. Use the variable x for this missing length.

Set up the proportion: $Tan(40°) = \frac{10}{x}$

Multiply both sides by x: $xTan(40°) = \frac{10}{x} \cdot x$

Simplify: $xTan(40°) = 10$

Divide both sides by the Tan(40°) $\frac{xTan(40°)}{Tan(40°)} = \frac{10}{Tan(40°)}$

Simplify, where Tan(40°) = 0.8391 $x = 11.92$

The correct answer choice is (B). Choice (A) is the length of RS. Choice (C) is the length of RT. If you chose choice (D), you used the Pythagorean Theorem, thinking that 40 was the length of the hypotenuse. Choice (E) would be incorrectly arrived at if you multiplied 100 by the Tan(40°).

Question 4: Finding a Missing Angle Using Trigonometry

What is the measure of angle ∠WXV below?

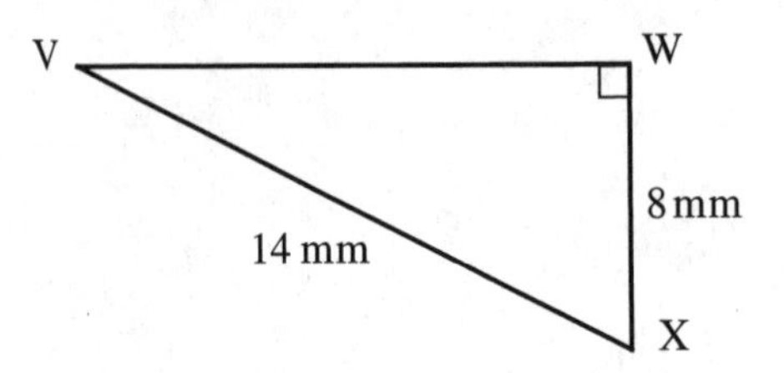

(A) 11.49°

(B) 22°

(C) 34.85°

(D) 55.15°

(E) 90°

In this triangle, you are given the length of the hypotenuse and the length of the side adjacent to angle ∠WXV. The ratio of the adjacent side to the hypotenuse is the cosine ratio. Use the $\cos^{-1}$ function to find the measure of the angle $\theta = Cos^{-1}\left(\frac{8}{14}\right)$. Using the calculator, $\theta = 55.15°$. **The correct answer choice is (D).** Choice (A) is the length of side VW. Choice (B) is the sum of the two given sides. Choice (C) is the measure of angle ∠WVX. Choice (E) is the measure of the right angle.

Question 5: Finding the Angle of Elevation

A surveyor is 16 feet from the base of a building. He sites the top of the building at an angle of 75°. How tall is the building?

(A) 4.29 feet

(B) 59.71 feet

(C) 61.82 feet

(D) 73.27 feet

(E) 91 feet

Draw the right triangle that represents this situation.

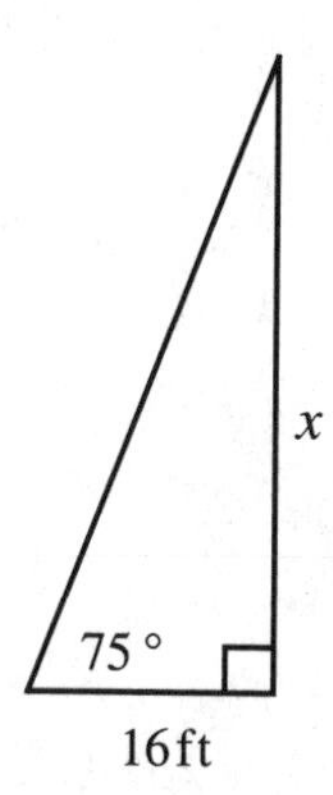

Because you know an angle and the adjacent side and you are asked to find the opposite side, the height of the building, use the tangent ratio.

Set up the proportion: $Tan(75°) = \frac{x}{16}$

Multiply both sides by 16: $16 \bullet Tan(75°) = \frac{x}{16} \cdot 16$

Simplify: $16(Tan(75°)) = x$

The $Tan(75°) = 3.732$, so: $16(3.732) = x$

$x = 59.71$

The correct answer is (B). If you chose (A), you set up the proportion incorrectly as $Tan(75°) = \frac{16}{x}$. Choice (C) is the length of the hypotenuse. Choice (D) is the result of treating 75 as the length of the hypotenuse and using the Pythagorean Theorem. Choice (E) is just the sum of 75 and 16.

CHAPTER QUIZ

1. Find the length of LN in the triangle below.

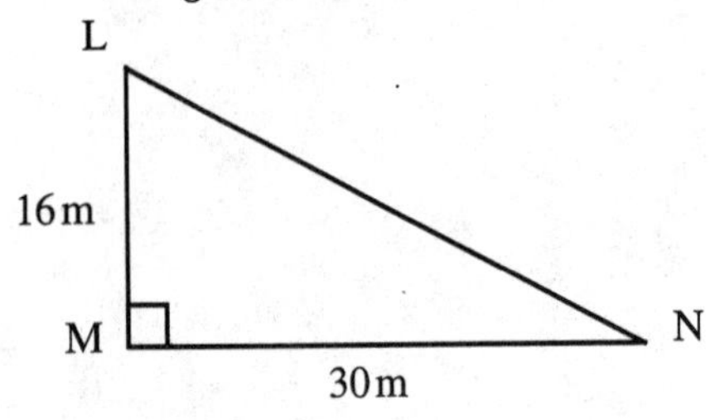

(A) 9.59 in

(B) 25.38 in

(C) 32 in

(D) 34 in

(E) 46 in

2. Which of the following side lengths of triangles is a right triangle?

(A) 2, 3, 4

(B) 3, 4, 6

(C) 10, 24, 26

(D) 10, 24, 30

(E) 9, 12, 21

3. The distance from River City to the State Park is 40 miles due south and 30 miles due east. What is the shortest distance between these two locations?

(A) 10 miles

(B) 11.84 miles

(C) 26.46 miles

(D) 50 miles

(E) 70 miles

4. What is the length of a leg of an isosceles right triangle with a hypotenuse of 8.49 inches?

(A) 4.25 inches

(B) 6 inches

(C) 8.49 inches

(D) 12 inches

(E) 14.7 inches

5. An equilateral triangle has sides that are 18 cm in length. What is the length of the altitude of this triangle?

(A) $9\sqrt{3}$ cm

(B) 18 cm

(C) cm

(D) 36 cm

(E) $36\sqrt{3}$ cm

6. What is the cosine of angle $\angle A$ in the triangle below?

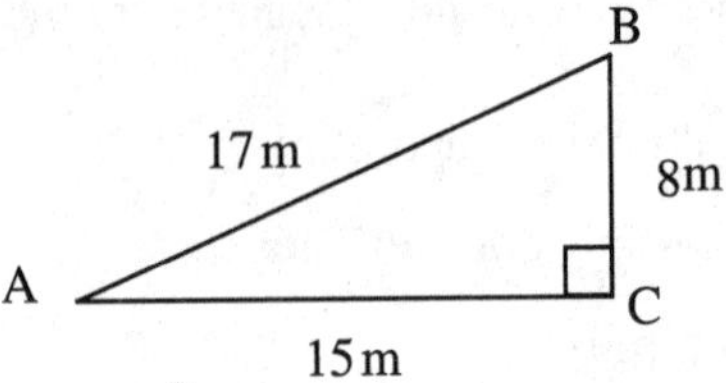

(A) $\frac{8}{17}$

(B) $\frac{8}{15}$

(C) $\frac{15}{17}$

(D) $\frac{15}{8}$

(E) $\frac{17}{15}$

7. For safety, it is recommended that the angle formed by the base of a ladder and the ground be no larger than 70°. Using this degree measure, what is the vertical height that a 30-foot ladder can reach?

   (A) 28.19 feet

   (B) 30 feet

   (C) 31.93 feet

   (D) 63.25 feet

   (E) 100 feet

8. What is the measure of angle $\angle Q$ in the triangle below?

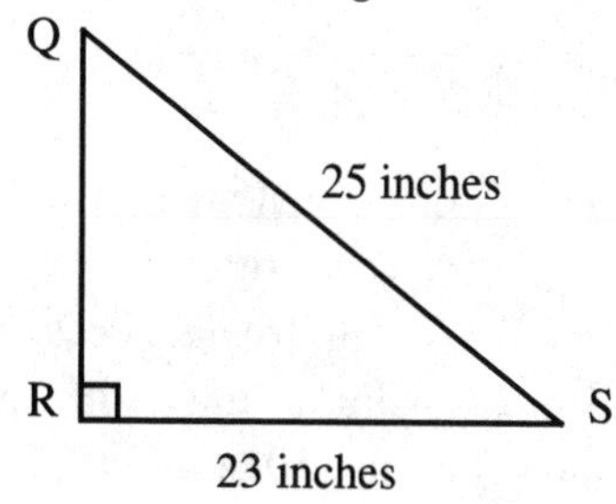

   (A) 0.016°

   (B) 23°

   (C) 25°

   (D) 43°

   (E) 67°

9. A sentry is at the top of a 50-foot lighthouse and sites a boat at an angle of depression of 45°. How far is the boat from the base of the lighthouse?

   (A) 25 feet

   (B) 35.36 feet

   (C) 50 feet

   (D) 70.71 feet

   (E) 86.6 feet

10. In the triangle below, what is the length of the altitude WY?

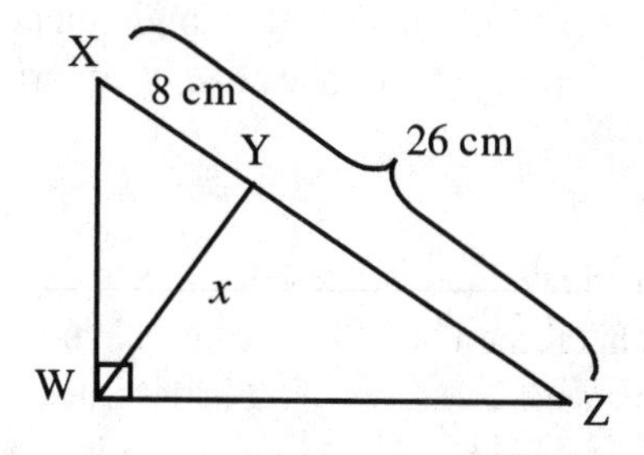

   (A) 12 cm

   (B) 13 cm

   (C) $8\sqrt{3}$ cm

   (D) 14.4 cm

   (E) 21.6 cm

ANSWER EXPLANATIONS

1. D

Use the Pythagorean Theorem:	$a^2 + b^2 = c^2$
Substitute into the theorem:	$16^2 + 30^2 = c^2$
Simplify:	$256 + 900 = c^2$
Add the terms on the left-hand side:	$1{,}156 = c^2$
Take the square root of both sides:	$\sqrt{1{,}156} = \sqrt{c^2}$
	$34 = c$

If you chose choice (A) you calculated $\sqrt{60 + 32}$, by multiplying by two instead of squaring. Choice (B) would be a leg of a right triangle with hypotenuse of 30. If your choice was (C), you may have thought this was a 30-60-90 degree triangle. In choice (E), you calculated 60 + 32, by multiplying by two instead of squaring, and then dividing by two.

2. C

You need to test these triplets to see which set will make the Pythagorean Theorem, $a^2 + b^2 = c^2$, true. It is not true that choice (A) is a right triangle because $2^2 + 3^2 \neq 4^2$. It is not true that choice (B) is a right triangle because $3^2 + 4^2 \neq 6^2$. For choice (C), it is true that $10^2 + 24^2 = 26^2$, so this is the right set. It is not true that choice (D) is a right triangle because $10^2 + 24^2 \neq 30^2$. It is not true that choice (E) is a right triangle because $9^2 + 12^2 \neq 21^2$.

3. D

The distance from the city to the park is described as due south and due east. This will form the legs of a right triangle, where $a = 40$ miles and $b = 30$ miles. The shortest distance will be the diagonal between the locations, the hypotenuse.

Use the Pythagorean Theorem:	$a^2 + b^2 = c^2$
Substitute into the theorem:	$40^2 + 30^2 = c^2$
Simplify:	$1{,}600 + 900 = c^2$
Add the terms on the left-hand side:	$2{,}500 = c^2$
Take the square root of both sides:	$\sqrt{2{,}500} = \sqrt{c^2}$
	$50 = \text{c}$

If your answer was choice (A), you subtracted the two distances. Choice (B) would be the result of multiplying the legs by two instead of squaring. If you chose (C), you thought that 40 miles was the hypotenuse and solved for a missing leg length. Choice (E) is the sum of the two measures, which is not the shortest distance between the locations.

4. B

You are told that this is a right isosceles triangle, so the hypotenuse is $\sqrt{2}$ times the length of the legs. Given that the hypotenuse is 8.49, divide by $\sqrt{2}$ to find the correct length. $8.49 \div \sqrt{2} = 6$ inches. If your choice was (A), you divided by 2 instead of $\sqrt{2}$. Choice (C) is the length of the hypotenuse. Choice (D) would be incorrectly arrived at if you multiplied by $\sqrt{2}$. Choice (E) is the hypotenuse of 8.49 multiplied by $\sqrt{3}$.

5. A

The altitude to an equilateral triangle forms two smaller 30-60-90 degree right triangles, where the hypotenuse is the side of the equilateral triangle that is 18 cm. The shorter leg of the right triangle is one-half of this hypotenuse, or 9 cm. The altitude, the longer leg of the right triangle, is thus 9cm in length. If you chose (B), you gave the length of the hypotenuse. Choice (C) would be incorrectly arrived at by multiplying the hypotenuse by $\sqrt{3}$, instead of the shorter leg. Choice (D) is twice the length of the hypotenuse.

6. C

The cosine ratio is the ratio of the adjacent length to the hypotenuse. Side AC is the adjacent side to angle $\angle A$, and side AB is the hypotenuse. Therefore, the cosine of angle $\angle A$ is $\frac{15}{17}$. Choice (A) is the sine of angle $\angle A$. Choice (B) is the tangent of angle $\angle A$. Choice (D) is the tangent of angle $\angle B$. Choice (E) is the reciprocal of the cosine function.

7. A

To understand this question, it helps to draw a right triangle to represent the situation, where the hypotenuse is the ladder.

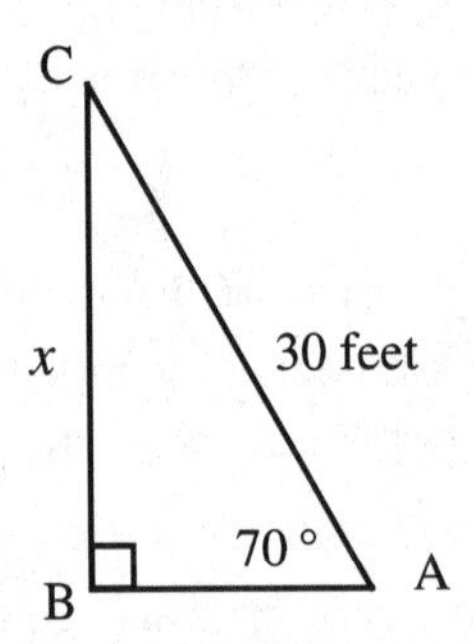

You are given angle ∠A of 70°, and the hypotenuse of length 30. You want to find the vertical height, that is side BC. Use the sine function, which compares the opposite side of angle ∠A to the hypotenuse.

Set up the proportion: $Sin(70°) = \frac{x}{30}$

Multiply both sides by 30: $30 \bullet Sin(70°) = \frac{x}{30} \cdot 30$

Simplify: $30(Sin(70°)) = x$

The Sin(70°) = 0.9397, so: $30(0.9397) = x$

$x = 28.19$ feet

If your answer was (B), you gave the height of the ladder, the hypotenuse of the right triangle. If your answer was (C), you divided 30 by the *Sin*(70°) instead of multiplying. Choice (D) would be an incorrect use of the Pythagorean Theorem, using 70 as a hypotenuse length and 30 as a length of a leg. Choice (E) is just the sum of 70 and 30.

8. E

You are given the length of the side opposite to angle ∠Q and the length of the hypotenuse. Use the sine function to find the missing angle measure.

$Sin(\theta) = \frac{23}{25}$, or $\theta = Sin^{-1}\left(\frac{23}{25}\right)$. Using your calculator, $\theta = 67°$, rounded to the nearest degree. If your choice was (A), you calculated the $\theta = Sin\left(\frac{23}{25}\right)$. Choice (B) is the measure of angle ∠S. Choice (C) is the length of the hypotenuse. Choice (D) would be the angle found by using tangent instead of sine.

9. C

Because the angle of depression is 45°, and this situation forms a right triangle, this is the special 45-45-90 degree right triangle, the right isosceles triangle. The height of the lighthouse is one of the congruent legs of the right triangle, so the distance of the boat to the lighthouse is also 50 feet. If you chose choice (A), you may have thought of the 30-60-90 degree triangle and that 50 was the hypotenuse. Choice (B) would be incorrectly calculated if you thought 50 was the hypotenuse and divided by $\sqrt{2}$. Choice (D) is the length of the hypotenuse. Choice (E) is equivalent to $50\sqrt{3}$.

10. A

You are given the length of the hypotenuse to triangle XWZ, and the length of the smaller segment of this hypotenuse. To solve this problem, use the mean proportional. First find the length of the longer segment of the hypotenuse. This will be 26 – 8 = 18 cm. Use the mean proportional, letting x represent the length of the altitude.

Set up the equation:	$x^2 = (8)(18)$
Multiply on the right-hand side:	$x^2 = 144$
Take the square root of each side:	$\sqrt{x^2} = \sqrt{144}$
	$x = 12$

Choice (B) is one-half of the hypotenuse. Choice (C) is $8\sqrt{3}$, so you may have been thinking of the 30-60-90 degree triangle. Choice (D) is side WX, not the altitude. Choice (E) is side WZ.